# CELLS & SYSTEMS
## *Living Machines*

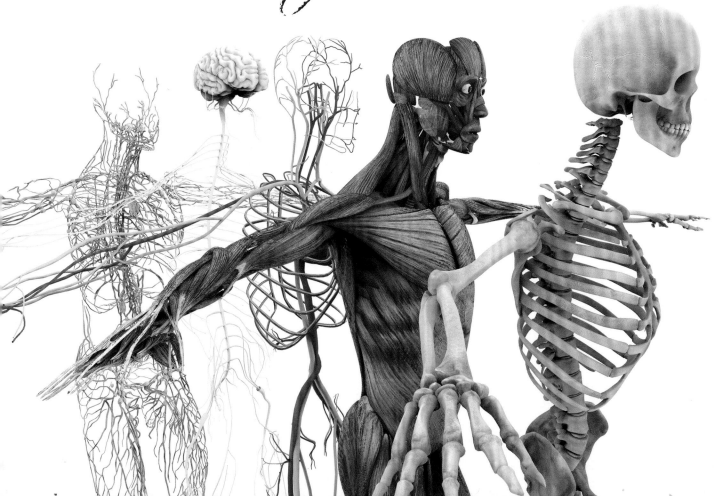

Dr. Heather Ayala is an Associate Professor of Biology and the Chair of the Department of Biology and Biochemistry at Belmont Abbey College. Over the past fifteen years she has worked at several small colleges throughout the country, including one year at Wyoming Catholic College. Dr. Ayala received her PhD in Biological Sciences from the University of Notre Dame, where she carried out research looking at the growth of *Plasmodium falciparum*, the parasite that causes malaria. Her primary interest is in teaching. At the college level, Dr. Ayala has taught courses in Parasitology, Anatomy & Physiology, Genetics, and many others. She has also taught Biology classes through co-ops for homeschool students at the high school level. She received an annual award at Belmont Abbey College in the spring of 2022 for her excellent teaching. Outside the classroom, Dr. Ayala enjoys spending time with her husband and four children, as well as cooking, sewing, gardening, and listening to good music.

MICROGRAPH (*PHOTOGRAPH TAKEN BY MEANS OF A MICROSCOPE*) OF BLOOD VESSEL

# CELLS & SYSTEMS
## *Living Machines*

Heather Ayala, PhD

TAN Books
Gastonia, North Carolina

*Cells & Systems: Living Machines* © 2022 Heather Ayala, PhD

All rights reserved. With the exception of short excerpts used in critical review, no part of this work may be reproduced, transmitted, or stored in any form whatsoever, without the prior written permission of the publisher. Creation, exploitation and distribution of any unauthorized editions of this work, in any format in existence now or in the future—including but not limited to text, audio, and video—is prohibited without the prior written permission of the publisher.

Unless otherwise noted, Scripture quotations are from the Revised Standard Version of the Bible—Second Catholic Edition (Ignatius Edition), copyright © 2006 National Council of the Churches of Christ in the United States of America. Used by permission. All rights reserved.

Excerpts from the English translation of the *Catechism of the Catholic Church* for use in the United States of America © 1994, United States Catholic Conference, Inc.—Libreria Editrice Vaticana. Used with permission.

Cover & interior design and typesetting by www.davidferrisdesign.com

ISBN: 978-1-5051-2624-2

Published in the United States by
TAN Books
PO Box 269
Gastonia, NC 28053

www.TANBooks.com

Printed in the United States of America

"Where were you when I laid the foundation of the earth?"

–Job 38:4

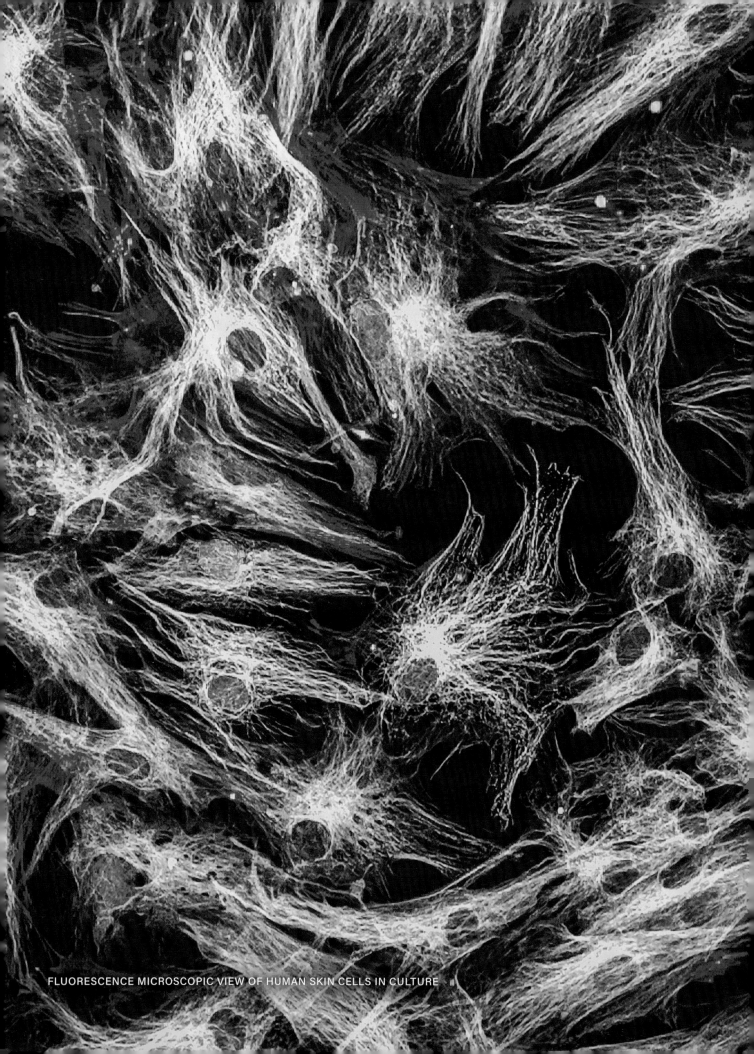
FLUORESCENCE MICROSCOPIC VIEW OF HUMAN SKIN CELLS IN CULTURE

# CONTENTS

Preface .................................................................................... **VIII**

Introduction ............................................................................ **XI**

Chapter 1: The Discovery of the Cell ................................... **1**

Chapter 2: A Tour of a Cell .................................................... **9**

Chapter 3: Making New Cells ................................................ **19**

Chapter 4: Tissues and Skin .................................................. **29**

Chapter 5: The Skeletal System ............................................ **37**

Chapter 6: The Muscular System .......................................... **49**

Chapter 7: The Nervous System ........................................... **59**

Chapter 8: The Cardiovascular System ................................ **71**

Chapter 9: The Respiratory System ...................................... **81**

Chapter 10: The Digestive System ........................................ **91**

Chapter 11: The Urinary System ........................................... **101**

Chapter 12: Inheritance & Human Development ................ **109**

Conclusion ............................................................................... **116**

Amazing Facts about Cells & Systems ................................. **117**

Key Terms ................................................................................ **124**

# PREFACE

When I think about the scientific study of the natural world, two phrases from the writings of Pope St. John Paul II come to mind:

(1) a rigorous pursuit of truth and

(2) a love of learning.

The first—a rigorous pursuit of truth—describes science and its processes. Scientists make careful observations, design experiments, and collect data to learn more about how the world works. Too often, though, science may seem like something you do in a big research facility with a lab coat.

But we are all scientists!

Anyone can study the living world in a scientific way. From an early age, everyone has a curiosity to understand the world. Think of a baby repeatedly dropping something onto the floor; they are discovering how gravity works! It is this basic curiosity that drives science.

The second piece—a love of learning—also describes what science should inspire. Sometimes science is depicted as a dry, boring set of facts, but nothing could be further from the truth. The world is a fascinating place. I have been interested in the natural world my whole life. This love of nature led me to obtain undergraduate and postgraduate degrees that have allowed me to teach biology classes every day for a living, and yet I am still constantly amazed by the wonders of our world.

There is always something new to learn in biology and all the natural sciences. Within biology, there is so much inspiring beauty and wonder in the endless forms continually changing and unfolding. The intricate order found in the smallest of God's creations—the cell—is so complex that it can help form and run the complexities of the physical life He has given us. The inner workings of the human body are truly awe-inspiring!

For example, did you know:

- If you stretched out a single chromosome, it would be about six feet long?

- A typical adult has 206 different bones in his body, from the largest, the femur in the leg just under twenty inches on average, to the smallest, at only a few millimeters inside our ears?

- A protein called hemoglobin found inside red blood cells is what gives blood its red color?

How could we not be fascinated by the cells and systems that make up our bodies? With this text, my friend and colleague, Dr. Heather Ayala, has done a marvelous job introducing you to this complex and intricate world, making it fun and edifying all at once. I have no doubt you will thoroughly enjoy this unit!

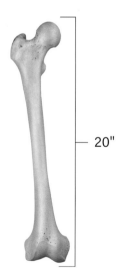

FEMUR BONE

— 20"

> *"[Science and faith] each can draw the other into a wider world, a world in which both can flourish."*
>
> —Pope St. John Paul II in *Physics, Philosophy and Theology*

Finally, in closing, it is too often assumed in our society today that faith and science act in opposition to one another, that somehow if we learn enough about the world, it would disprove the existence of God. But it is important for each of us to be confident in our Faith and the fact that truth cannot be in opposition with itself.

We read in the *Catechism of the Catholic Church*: "Methodical research in all branches of knowledge, provided it is carried out in a truly scientific manner and does not override moral laws, can never conflict with the faith, because the things of the world and the things of faith derive from the same God. The humble and persevering investigator of the secrets of nature is being led, as it were, by the hand of God in spite of himself, for it is God, the conserver of all things, who made them what they are" (*CCC* 159).

Holy Mother Church teaches us that we can pursue scientific knowledge unafraid. It is my hope that *The Foundations of Science* series will not simply give your family some facts about the world but also instill a curiosity and love of learning in you that you can apply across all the disciplines of your life, both scientific and otherwise.

**Timothy Polnaszek, PhD**

RED BLOOD CELLS

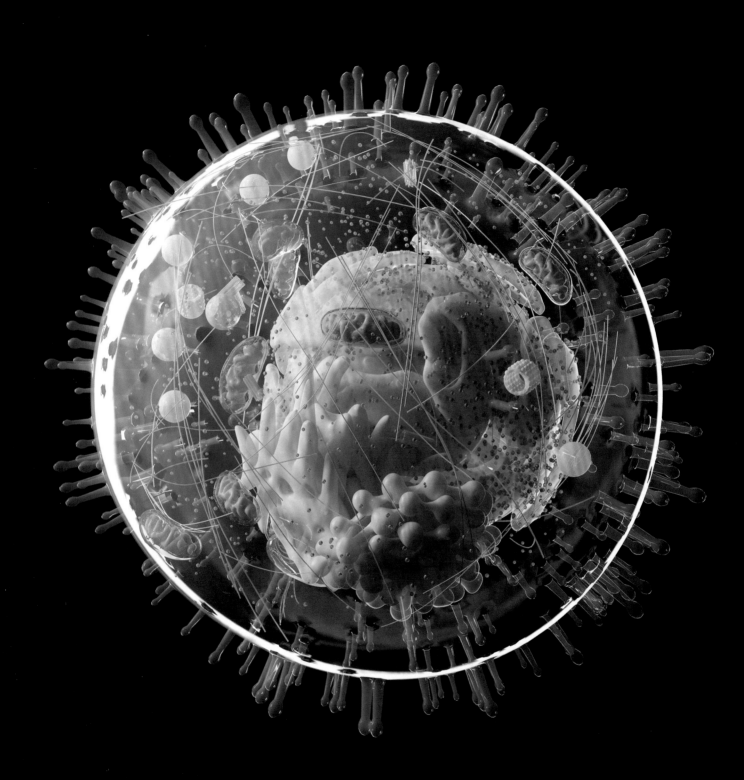

HUMAN CELL ANATOMY

# INTRODUCTION

So far in *The Foundations of Science* curriculum, we have explored many different things found in our world. Some of them have been living organisms, like plants and animals, while others have been non-living, like when we learned about the earth, the skies, and outer space. But all of the things we have explored so far, we have been able to see with our eyes.

That's all about to change!

All living things are made up of very small (so small you cannot see them!) units called cells. These **cells** are the basic building blocks of life; they make up all living organisms and the tissues of our bodies. You can think of cells kind of like LEGO bricks. They are different, but similar, and you can put many of them together to make an infinite number of different structures. Even though you cannot see them with the naked eye, there is an incredible amount of activity that goes on inside of every single cell. The structure of a cell is so intricately made that it can perform the functions needed for life to exist. We will explore some of these special roles in the first few chapters of our book.

Then, we will move on to tissues. When cells "work together" for a common purpose, they make up tissues. These tissues are the fabric used to create the organs such as your stomach, brain, heart, muscles, and bones. The remainder of this book will then look at many of the organs and organ systems that come together to make up your own body. By the time our journey is complete, I hope you will have a greater appreciation of the care God took in creating you!

Our study of cells dates all the way back to the 1600s, when Robert Hooke developed his own microscope to study ants, fleas, plant material, and many other things. He would pen a book called *Micrographia* that contained drawings like this one of what he observed.

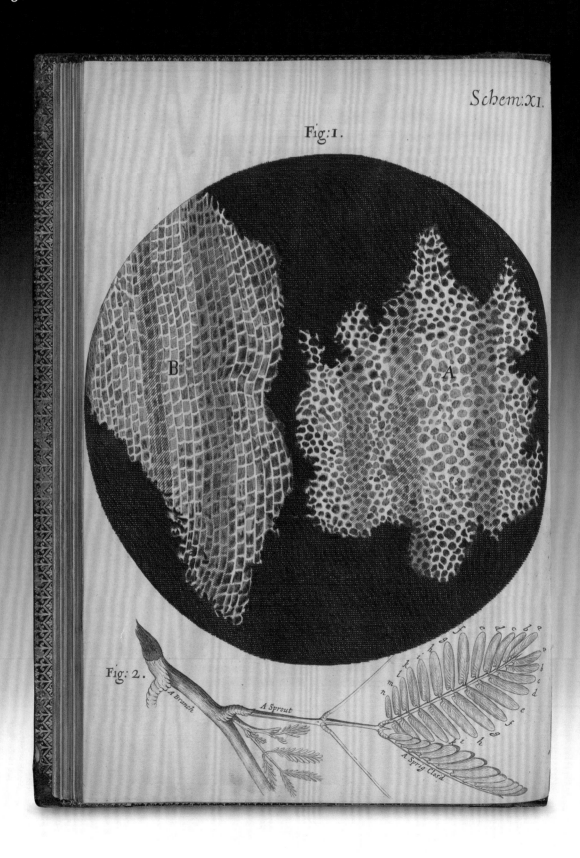

# CHAPTER 1

## THE DISCOVERY OF THE CELL

*The Dutch father and son Hans and Zacharias Janssen, in the late sixteenth century, were the first to develop the microscope. This invention, refined through the centuries, has expanded our knowledge of the world around us and the human body in unimaginable ways.*

## HOOKE'S MICROSCOPE

Have you ever used a magnifying glass to look at something small? Perhaps you used one to look at the fingerprints on your hand or at the veins on a leaf. If so, you know that by using the magnifying glass, it makes the thing you are looking at seem bigger. By doing this, it lets you see things that you may not have seen if you had not used the magnifying glass.

A **magnifying glass** is a lens, similar to the lens you would find in a pair of eyeglasses. The lens is a curved piece of glass. Because it is curved, when light passes through the lens, it causes the light to bend just a little bit. The bending of the light changes how we see the object on the other side of the lens, making it look bigger.

In the late 1500s, there lived a Dutch man by the name of Hans Janssen who made eyeglasses. Hans had a son named Zacharias. In 1590, Hans and Zacharias started to experiment with lenses, putting two lenses together in a tube. They found that when they looked at something through the tube of lenses, it looked even bigger! This was the first microscope.

We don't know a lot about Hans and Zacharias, but they are given credit for inventing the compound microscope. A **compound microscope** is an instrument that uses more than one lens to make objects look bigger than they really are. The invention of microscopes opened up a whole new world to people. Now scientists could look at small things that they had never seen before.

## How Do Microscopes Work?

Most microscopes you see today are compound light microscopes, meaning that they use two or more lenses and light. The light passes through a very thin object on a slide and is directed through the lenses. As it passes through the lenses, the light bends before it reaches your eye. This produces an image that looks bigger than it actually is. The lenses used in a compound light microscope are called **converging lenses**, like the one pictured here. Notice that the image formed is not only bigger but also upside down and backwards from the original object. This means that when you look through a microscope, if you move the slide to the right, you will see the image move to the left. And if you move the slide up, you will see the image move down!

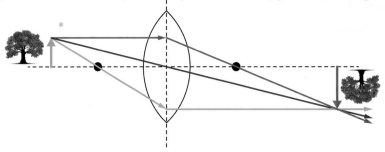

Unfortunately, microscopes were expensive, and most people never had a chance to see one or look through one. But in 1665, about seventy years after the microscope was invented, an English scientist by the name of Robert Hooke made his own microscope. He had a practice of observing ants, fleas, plant material, and many other things. Hooke made detailed drawings and descriptions of what he saw, and he published all of these together in a book called *Micrographia.* In one part of his book, Hooke describes his observations of an ant. In order to make these observations, he first had to find a way to keep the ant from moving around. If he killed the ant, then it would be squished, and he would not be able to see its proper shape. He tried to use glue to hold the ant in place, but the ant kept wiggling about so that Hooke could not observe it under the microscope. Finally, Hooke put the ant in some strong alcohol to put it to sleep. This worked for about an hour, but then the ant suddenly woke up and started to run away before Hooke could finish his observations! The people found Hooke's book to be very interesting to read and fun to look at because of the beautiful drawings. This made it a very popular book.

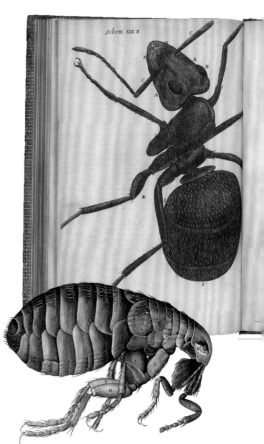

Another of Hooke's famous sketches was of dead plant material, called cork. Hooke cut a thin slice of a piece of cork and looked at it under his microscope. He saw that the cork looked like little boxes, or pores, like what you might find in a honeycomb. Hooke called these boxes cells. This is where we get the word for cells!

## BASIC CELL STRUCTURE

When Hooke made his sketch and description of the cells in cork, he was observing the shell, or outer walls of a plant cell (the inside of the cell was no longer present). This outer wall gave the cork a boxlike or square appearance, but not all cells contain an outer wall. Animal cells (which include our cells, since we humans are animals too) do not have a cell wall. As a result, they come in many different shapes and sizes. We will discuss more about the different types of cells in chapter 2. For now, let's focus on some of the things that are common to *all* cells.

Every cell has an outer skin of sorts (think like the peel of an apple) that holds it together. This is called the **cell membrane** (sometimes also called the **plasma membrane**). The cell membrane helps control what can move in and out of the cell. Inside of living cells is a gel-like fluid made mostly of water. We call this the **cytoplasm** of the cell. Floating in the cytoplasm are other small structures. We call these structures **organelles**, or "little organs." We will talk about these in more detail in chapter 2.

The last thing we find in every cell is genetic material in the form of **DNA (Deoxyribonucleic acid)**. This genetic material contains the "instructions" that tell the cell what organelles to build and how to carry out its job. DNA is like the directions that make you who you are (how tall you are, the color of your skin, etc.), like blueprints for how to build a house.

# BACTERIAL CELL ANATOMY

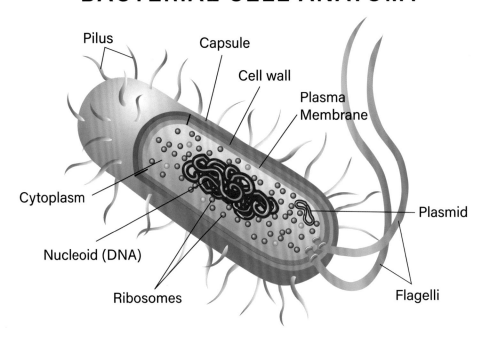

## DEVELOPMENT OF THE CELL THEORY

Scientists continued to improve on the microscope in the years following its invention. This allowed them to make discoveries that helped increase our understanding of the cell and its role in making up living organisms. There were several very important observations made about cells during this time. We will just mention a few of them here.

- Henri Dutrochet was a French military doctor living in the early nineteenth century. In examining plant cells, he noted that cells are the basic structural component making up plants. He also observed that the functions carried out by the plant cells are the basis for the function of the plant.

### How Big Is the Biggest Cell?

Most cells are too small to be seen without a microscope, but there are some cells that are bigger than others. For example, did you know that an egg is a single cell? The biggest egg is laid by the ostrich. It is about six inches long and weighs about three pounds! While this may be the *heaviest* cell, it is not the *biggest* cell. One type of cell we will learn more about in chapter 7 is a nerve cell. Nerve cells make up our brain and other nervous structures throughout our body. Their job is to carry messages from one part of our body to another. We have nerves stretching from our spine to the bottom of our toe. That is a long nerve cell. But giraffes have even longer nerve cells. They have nerves that run the entire length of their long necks, between six and eight feet long! However, the longest nerve cell is found in the giant squid. These sea animals can be up to forty-three feet long and have nerves up to thirty-nine feet long. Now that is a big cell!

- In 1832, the Belgian scientist Barthelemy Dumortier was making observations when he saw that a plant cell divides into two new plant cells. He concluded that all new cells come from pre-existing cells.

- In the years 1838–39, two German scientists named Schwann and Schleiden were studying a variety of plant and animal cells. Schleiden saw that all of a plant's structure is composed of plant cells. Schwann's work, meanwhile, focused on studying animals. Like Schleiden, he noted that all the structural parts of animals are made of animal cells. In short, Schwann concluded that plants and animals are similar in that they are both made of up of cells and these cells work in the same basic way.

If that seems like a lot to remember, don't worry. The most important thing to understand is that all of these discoveries combine to form what we call the **Cell Theory**. The Cell Theory says:

1. All living things are made of cells.
2. Cells are the basic unit of structure and function for all living things.
3. All cells come from pre-existing cells.

These ideas form the foundation for our current understanding of cells and their role in making up all living things. In chapter 2, we will look more closely at different types of cells and what makes up these small building blocks of life.

*Remember:*
*The DNA (Deoxyribonucleic acid) found in cells is like the "directions" that make you who you are (physically speaking).*

## FOUNDATIONS REVIEW

✓ The invention of the magnifying glass, which enlarges what you are looking at, and more specifically the compound microscope, helped scientists see things that otherwise would have been too small. This led to the discovery of cells, the basic building blocks of life.

✓ Though different types of cells possess different qualities, every cell has an outer skin that holds it together. This is called the cell membrane. Also common to all cells is the cytoplasm, the gel-like fluid made mostly of water, and floating in this cytoplasm are little structures called organelles, or "little organs." The last thing we find in every cell is genetic material in the form of DNA (Deoxyribonucleic acid). This genetic material contains the "instructions" that tell the cell what organelles to build and how to carry out its job.

✓ Across many, many years of research, scientists from all over the world contributed to the development of what is known as the Cell Theory. It states that: (1) All living things are made of cells; (2) Cells are the basic unit of structure and function for all living things; (3) All cells come from pre-existing cells.

# Disproving Spontaneous Generation

The idea that cells come from pre-existing cells was astounding in 1750. Prior to the 1700s, many people believed that small, simple living things like worms and maggots (fly larvae) could just spontaneously arise out of nonliving material. This idea was known as "spontaneous generation." Does that sound strange to you?

One of the earliest records we have of the theory of spontaneous generation comes from the Greek philosopher Aristotle. He lived about 350 years before Christ was born. Aristotle suggested that life could come from nonliving things. Aristotle observed a pond that did not appear to have any living animals, and yet within a few days or weeks, fish could be seen swimming about. This led him to believe that the fish spontaneously appeared from the water itself!

Of course, we now understand that the fish came from eggs that were in the pond. But other scientists made similar observations over the years that seemed to support Aristotle's idea. They would observe frogs "appearing" along the banks of the Nile River. Really, the frogs were developing from tadpoles that were already in the water. Another observation noted that rotting meat left out would soon start to have maggots growing on it!

This idea of spontaneous generation was very popular for hundreds of years. It was not until the 1600s that scientists started to question it. The first person to perform an experiment that showed spontaneous generation to be wrong was a Catholic Italian doctor named Francesco Redi in 1668. Redi lived towards the end of the time of Galileo. He was inspired by Galileo's approach of gathering observations to answer a scientific question. Redi decided that he would use a similar approach to investigate the idea of spontaneous generation.

Redi developed an experiment to test if maggots spontaneously appeared on rotting meat. He put some rotting meat into several different containers. Some of these containers he left open to the air. A second set of containers were sealed. A third set of containers were covered with gauze. This allowed air to move in and out of the container, but no flies could get to the meat. After some time, Redi observed that the uncovered meat had maggots growing on it, but the meat in the sealed container and the meat in the gauze-covered container did not. Redi observed maggots hatching from eggs that were laid on the gauze covering but not on the meat within the gauze-covered container. Based on his observations, Redi concluded that the maggots were not arising out of the meat but that they were hatching from the eggs laid by the flies on the meat.

Redi's experiment was the first of many performed in the 1600s and 1700s that showed the theory of spontaneous generation was incorrect. His work was also very important because it introduced the idea of a **controlled experiment,** which is when multiple experiments are carried out with one variable or factor being changed in order to observe the effects of the change in variable. This is very important in how we answer scientific questions today.

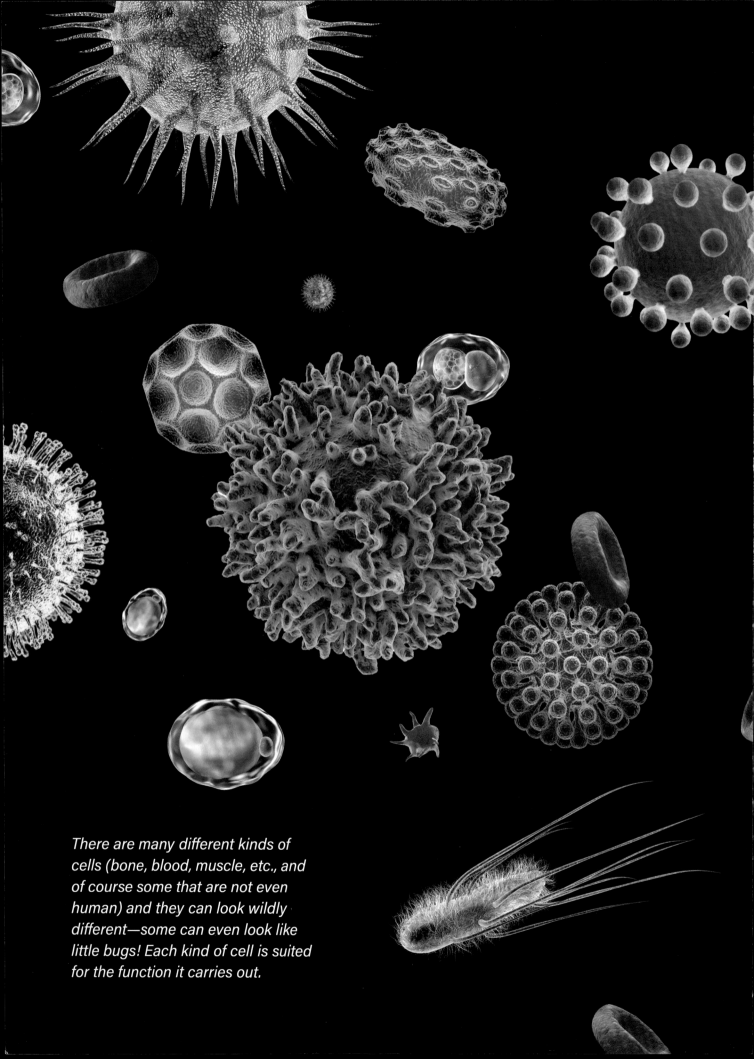

*There are many different kinds of cells (bone, blood, muscle, etc., and of course some that are not even human) and they can look wildly different—some can even look like little bugs! Each kind of cell is suited for the function it carries out.*

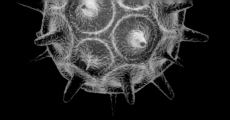

# CHAPTER 2

## A TOUR OF A CELL

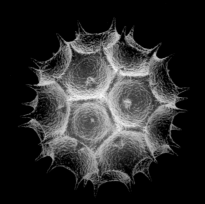

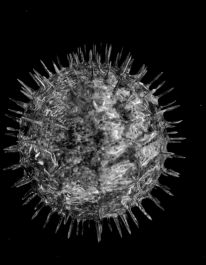
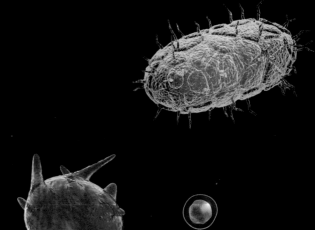

# THE CELL MEMBRANE

In chapter 1, we learned about how microscopes led to the discovery of the cell, and we were introduced to the three main components found in all cells: cell membrane, cytoplasm, and genetic material (DNA). Before we explore the cell in more detail, let's spend some time learning about the specific structure and role of the cell membrane.

As we mentioned already, the cell membrane is the outer "skin" of the cell. It gives a cell its shape. It also helps control what can enter and exit the cell. You could think of the cell membrane kind of like the wall around a medieval city. It protects the city and controls what can enter and exit through its gates. Another term for the cell membrane is the plasma membrane.

The structure of the cell membrane is made mostly of special molecules called **phospholipids**. These molecules are made of a "head" and two "tails." The head has chemical properties that allow it to interact well with water. We say that it is **hydrophilic**, or water-loving. The second part of the phospholipid is the two tails; these are **hydrophobic**, or water-fearing ("phobia" means to fear something). This means the tails have chemical properties that cause them to repel water. Fats and oils are examples of hydrophobic substances. Do you know what happens when you mix water with oil? The water and oil separate, with the oil floating on the water. This shows us that hydrophilic and hydrophobic molecules do not like to mix with each other (do not "attract" each other), since one repels water and the other is attracted to it.

The phospholipid molecule is very special because it is both hydrophilic and hydrophobic. If you were to take thousands of phospholipid molecules and put them into a cup of water, they would arrange themselves into two rows. Biologists call this a "phospholipid bilayer" because there are two layers of phospholipids. The heads would be on the top and bottom, close to the water. In the middle would be the long hydrophobic tails hiding away from the water. Perhaps it is helpful to think about this using an analogy. Let's say that the two layers of phospholipids are arranged like an Oreo cookie. The two chocolate cookie wafers are on the top and bottom like the heads, while the tails are the creamy frosting in the middle.

The special arrangement of the cell membrane controls what can move into or out of the cell. Molecules that are hydrophilic or have a charge cannot cross the cell membrane. Large molecules also cannot cross the cell membrane. But just like a medieval city needs to bring food in from the farms to feed the people, a cell also needs to bring in food particles, which can be rather large (at least relatively speaking). This can be done using **proteins**, which are large, complex molecules that play many critical and beneficial roles for our bodies (you want to eat a diet that is high in protein). Cell membranes have proteins that are stuck into the membrane. These work like tunnels through that city wall, allowing the food molecules to move into the cell.

*Remember:*
*Phospholipids are made of two parts: a head, which interacts well with water, and two tails, which repel water. This unique structure plays an important role in what moves in and out of the cell.*

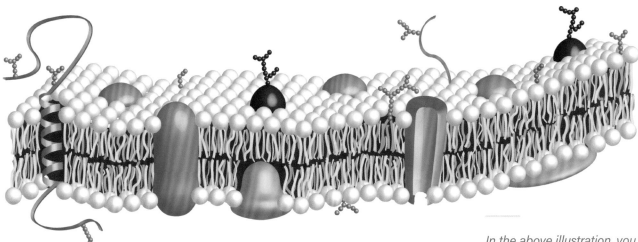

## PROKARYOTIC AND EUKARYOTIC CELLS

While all cells have cell membranes, not all cells look or act the same. Scientists have identified characteristics of cells and put them into different groups based on what they look like and what they do, sort of like if a teacher organized her class into groups based on height or hair color or the favorite hobbies of the students. The first factor used to group cells is where the DNA is, either enclosed in a membrane *inside* the cell or "floating" freely in the cytoplasm. Cells that have free-floating DNA are called **prokaryotic cells**, while cells that have their DNA enclosed in a membrane structure are called **eukaryotic cells**.

This is not the only difference between prokaryotic and eukaryotic cells though. Prokaryotes are generally much smaller and simpler than eukaryotes. An example of a prokaryote is a bacteria cell. You've probably heard the word "bacteria" before. **Bacteria** are small organisms (living things) made of a single cell, unseen by the naked eye. Some bacteria can make you sick (like germs), but some can actually be good for you.

Bacteria cells have a cell membrane, cytoplasm, and DNA, but they often have some other structures as well. **Ribosomes** are small structures inside the cell that make proteins. Proteins carry out many important cell functions. One we already mentioned is that proteins allow substances to pass in and out of cells through the cell membrane (they form the "tunnel"). Other proteins perform special functions inside our cells, like breaking down food. And proteins are also very important building blocks for making other structures in cells and in our bodies, such as muscle and hair, which we will talk about in later chapters.

*In the above illustration, you can see a depiction of how the phospholipids arrange themselves into a phospholipid bilayer and why our analogy to an Oreo cookie makes sense. The heads on the top and bottom would be the chocolate wafers, while the tails in the middle are the cream. You can see also how the proteins (light blue) create passageways through the membrane.*

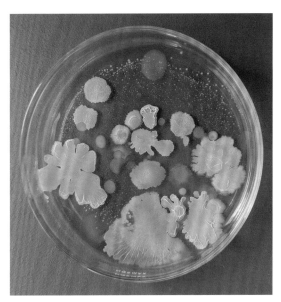

BACTERIA COLONIES

## What Does It Mean to Be Lactose Intolerant?

One of the roles of proteins in cells is to help break down larger molecules into smaller molecules, like taking apart LEGO bricks. We call these special proteins **enzymes**. Let's look at an example. The enzyme lactase helps to break down the sugar lactose into two smaller sugar molecules. Lactose is found in milk and other dairy products like cheese. The lactose fits into a special hole in the lactase enzyme. This is similar to how a key fits into a keyhole. Maybe you have heard of someone who is lactose intolerant. This means that if he drinks milk, his body cannot break down the lactose, and as a result, he can feel sick. Lactose intolerance occurs because the person's body is not making enough of the lactase enzyme. Lactase is just one example of an enzyme, but there are thousands of different enzymes in our cells, each with its own specific shape and function.

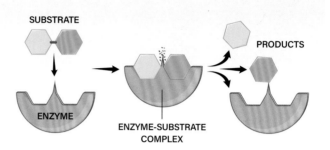

*Cells Fun Fact:*
*Human cells have twenty-three pairs, or forty-six chromosomes. If you stretched out a single chromosome, it would be about six feet long!*

CHROMOSOME

Another structure that you might find in bacteria cells is a **cell wall**. The cell wall is found outside of the cell membrane. It is made of proteins and other molecules that help to protect the cell from the outside environment. Some bacteria may have a tail-like structure called a **flagellum**. The flagellum of a prokaryote rotates or spins. This helps the bacteria cell to move, just like a tail helps a fish swim or a propeller pushes a boat through the water.

In contrast to prokaryotic cells, eukaryote cells are larger and more complex. The **nucleus** of a cell sits at the center of the cell. You might compare it to our heart, in the center of our bodies. The nucleus in a eukaryotic cell is a membrane-bound organelle that contains the DNA. The nuclear membrane has the same structure as the membrane surrounding the cell. However, the nucleus is surrounded by *two* plasma membranes. This helps to protect the contents of the nucleus. As we mentioned in chapter 1, the DNA contains all of the information a cell needs to carry out its specific functions. If some of that information goes missing or gets damaged, that would be very bad for the cell, and for the overall organism. You can see why it is very important, then, to protect the DNA inside the nuclear membrane. The DNA in the nucleus is wrapped around proteins to form **chromosomes**. This is like wrapping thread around a spool and helps to keep the DNA organized in the nucleus.

In addition to a membrane-bound nucleus, eukaryotic cells have other membrane-bound structures and organelles. One of these is called the **mitochondria**. These are small bean-shaped structures enclosed by two plasma membranes. The purpose of the mitochondria is to break down sugars and mine energy for the cell. Just like you need to eat food so that you can run, play, and grow, cells also need energy to be able to carry out special functions. Inside the mitochondria are many special enzymes that carry out chemical reactions to break down sugar molecules into energy for the cell through a process called **cellular respiration**. For this reason, the mitochondria are

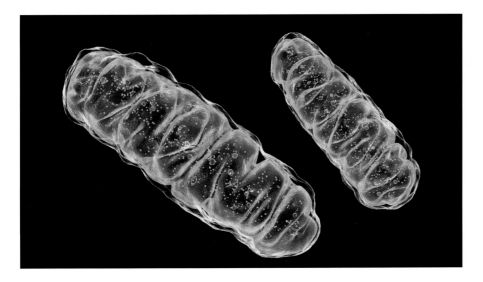

*This 3D depiction of the bean-shaped mitochondria shows it wrapped in plasma membranes with enzymes inside. The mitochondria is often referred to as the powerhouse of the cell since it breaks down sugar molecules to create energy.*

sometimes called the powerhouse of the cell. You could think of them like the power plant that burns coal to produce the electricity that gets sent to your house.

## PLANT VERSUS ANIMAL CELLS

Eukaryotic cells include both plant and animal cells. There are some important differences between plant and animal cells. Plant cells have a cell wall and animal cells do not. The cell wall of a plant is made differently than the cell wall of bacteria. Plant cell walls are made of a complex sugar molecule called **cellulose**. The cell wall gives plants support so that they can grow tall. It also helps protect plants from other things that might try to harm them.

Another difference between plant and animal cells is that plants have a special organelle called a **chloroplast**. You learned about chloroplasts in our *Plants* book. Remember that chloroplasts are small organelles that capture light energy from the sun. They use the light energy to transform carbon dioxide gas

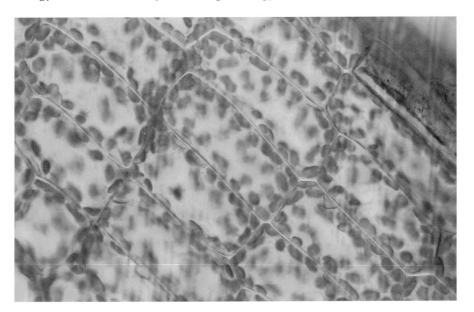

*This magnified image (400x) of dotted thyme-moss allows us to see the structure of plant cells and the chloroplasts that help plants create energy through the process of photosynthesis.*

and water into chemical energy through the process of photosynthesis. This is how plants make their own "food." You can think of the chloroplast kind of like a solar panel. It collects light from the sun and transforms it into energy that the plant can use.

### CELL ORGANELLES AND THEIR FUNCTIONS

| Cell Organelle | Description | Function |
| --- | --- | --- |
| Cell (plasma) membrane | Two layers of phospholipids | Protect the cell, controls what can move in and out of a cell |
| Ribosomes | Very small protein structures | Make proteins |
| Nucleus | Core of the cell, enclosed in a plasma membrane | Contains DNA in the form of chromosomes |
| Mitochondria | Bean shaped organelle, enclosed in a plasma membrane | Produces energy for the cell |
| Chloroplasts | Disc shaped, green organelle, enclosed in a plasma membrane | Captures light energy from the sun and changes it into food for plants. |

## ANATOMY OF A CELL

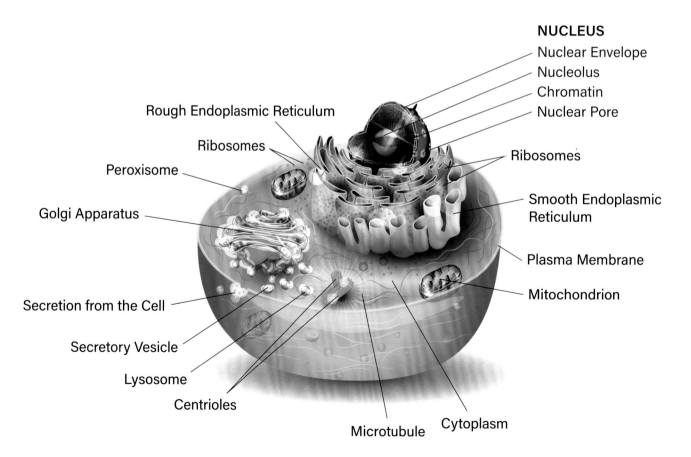

## THE COMPLICATED WORLD OF CELLS

In this chapter, we learned about some of the differences between prokaryotic cells and eukaryotic cells. We also explored some of the organelles found in cells, including the nucleus, ribosomes, mitochondria, and chloroplasts. There are many other organelles found inside of cells, but unfortunately, we don't have time to talk about all of them. The world of cells is very complicated, as I'm sure you are now realizing! We spoke about so much, and some of it may have been difficult to understand. But for now we just want to understand the basics.

If you are interested, you could do some extra reading on your own and learn about other organelles in different cells. Isn't it amazing how many structures are found in each of your cells? All of them are necessary for making our cells work properly. In our next chapter, we will turn to some of the other important functions our cells carry out.

### FOUNDATIONS REVIEW

✓ The structure of the cell membrane is made mostly of special molecules called phospholipids. These molecules are made of a "head" and two "tails." The head has chemical properties that allow it to interact well with water. We say that it is hydrophilic, or water-loving. The second part of the phospholipid is the two tails; these are hydrophobic, or water-fearing. This means the tails have chemical properties that cause them to repel water. These properties of the cell serve important functions, such as helping to determine what goes in and out of the cell.

✓ While all cells have cell membranes, not all cells look or act the same. Scientists have identified characteristics of cells and put them into different groups based on what they look like and what they do. Two important categories of cells are prokaryotic cells and eukaryotic cells. Prokaryotic cells tend to be smaller and less complex and their DNA "floats" in the cytoplasm, while eukaryotic cells are bigger, more complex, and their DNA is enclosed inside a membrane structure called a nucleus.

✓ Eukaryotic cells include both plant and animal cells. There are some important differences between plant and animal cells. For example, plant cells have a cell wall and animal cells do not. Additionally, plants have special organelles called chloroplasts in their cells, which are small organelles that capture light energy from the sun, allowing plants to make their own food through photosynthesis. This is something animals cannot do (we must go out and find our food).

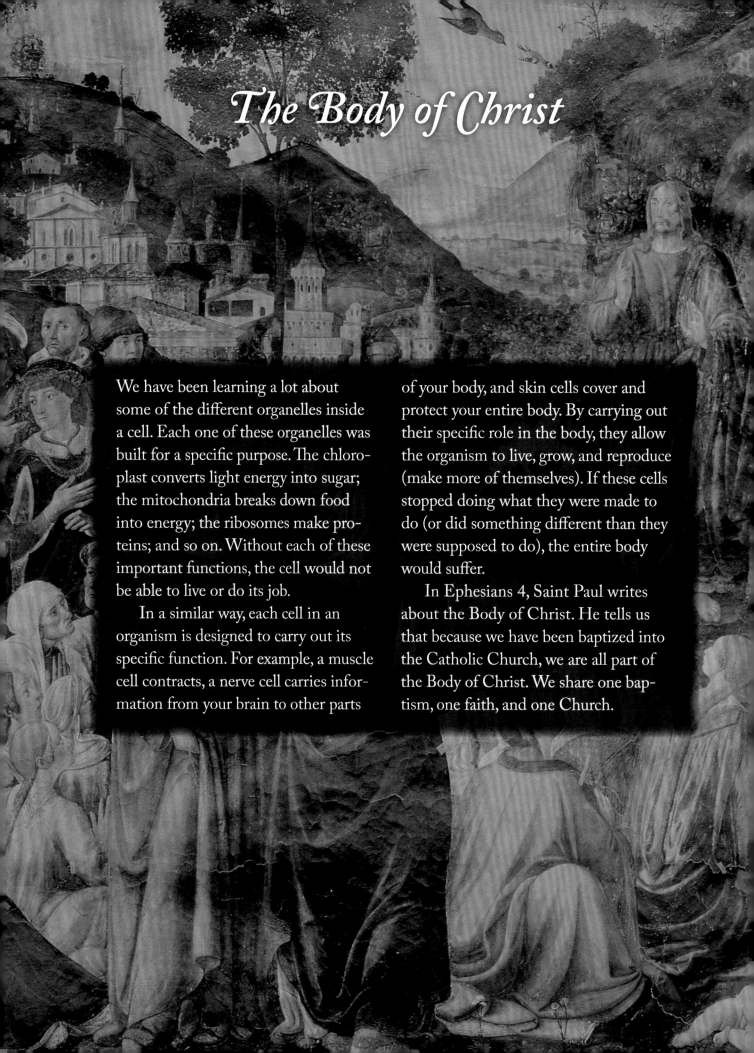

# The Body of Christ

We have been learning a lot about some of the different organelles inside a cell. Each one of these organelles was built for a specific purpose. The chloroplast converts light energy into sugar; the mitochondria breaks down food into energy; the ribosomes make proteins; and so on. Without each of these important functions, the cell would not be able to live or do its job.

In a similar way, each cell in an organism is designed to carry out its specific function. For example, a muscle cell contracts, a nerve cell carries information from your brain to other parts of your body, and skin cells cover and protect your entire body. By carrying out their specific role in the body, they allow the organism to live, grow, and reproduce (make more of themselves). If these cells stopped doing what they were made to do (or did something different than they were supposed to do), the entire body would suffer.

In Ephesians 4, Saint Paul writes about the Body of Christ. He tells us that because we have been baptized into the Catholic Church, we are all part of the Body of Christ. We share one baptism, one faith, and one Church.

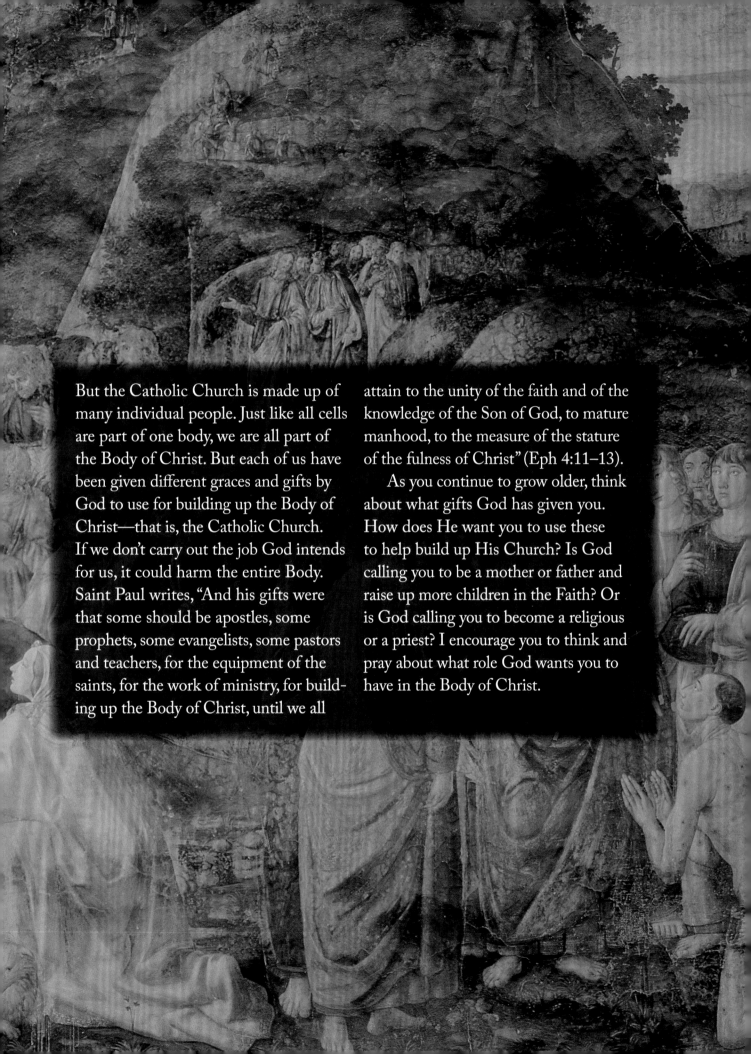

But the Catholic Church is made up of many individual people. Just like all cells are part of one body, we are all part of the Body of Christ. But each of us have been given different graces and gifts by God to use for building up the Body of Christ—that is, the Catholic Church. If we don't carry out the job God intends for us, it could harm the entire Body. Saint Paul writes, "And his gifts were that some should be apostles, some prophets, some evangelists, some pastors and teachers, for the equipment of the saints, for the work of ministry, for building up the Body of Christ, until we all attain to the unity of the faith and of the knowledge of the Son of God, to mature manhood, to the measure of the stature of the fulness of Christ" (Eph 4:11–13).

As you continue to grow older, think about what gifts God has given you. How does He want you to use these to help build up His Church? Is God calling you to be a mother or father and raise up more children in the Faith? Or is God calling you to become a religious or a priest? I encourage you to think and pray about what role God wants you to have in the Body of Christ.

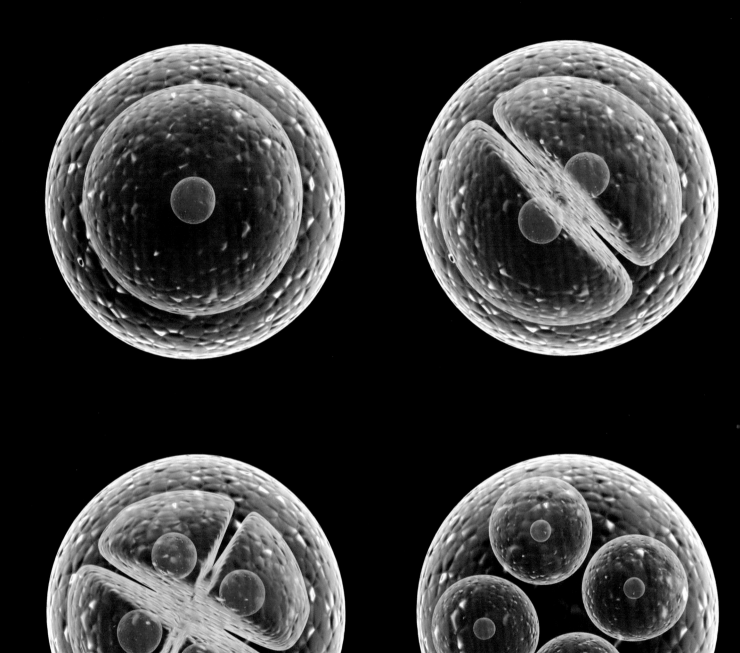

*The ability to reproduce is key to the survival of any living species, and that process is dependent upon the process of cell division.*

# CHAPTER 3

*MAKING NEW CELLS*

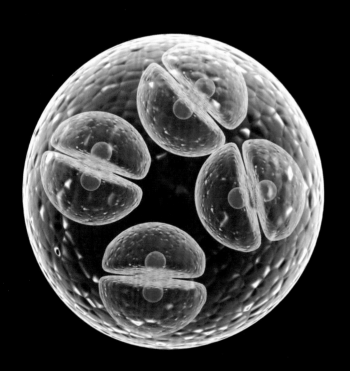
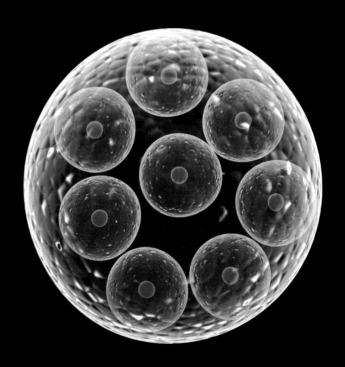

## ...AND MULTIPLY

In the first chapter of Genesis, we read about how God created the world, how He created the light and darkness, day and night, land and sea, plants, animals, and man. But something you may not have noticed is that after God creates the birds and sea creatures and animals, He commands them to be fruitful and multiply. They would not be able to do this if God had not given them a way to make more of themselves (**reproduce**). The ability to reproduce is a key characteristic to all living things. If living things could not reproduce, then life on Earth would not be able to continue to exist.

## CELL DIVISION

In chapter 1, we learned that cells come from preexisting cells. This means that cells are also able to make more of themselves. We call this process **cell division**. Cell division is necessary for many reasons. If you fall and get a cut on your knee, your body has to make new cells (new skin if you want to think of it that way) in order to heal the cut. Another important need for cell division is to help you grow. You get bigger and taller because your body keeps adding more cells. For some organisms that are made of only a single cell (like a bacteria cell), cell division helps them to make more of their kind.

Eukaryotic cells reproduce through a process called **mitosis** (pronounced *my-toe-sis*). The first thing that must happen before a cell divides is that it must make a copy of all of its DNA. Each new cell will need a full copy of the DNA in order to function properly. The cell makes a perfect copy of the DNA in a process called **DNA replication**. Here's a way to consider it: Let's say that you want to make some cookies. To do so, you would follow the instructions on a recipe. You share your cookies with a friend, and they want the recipe so they

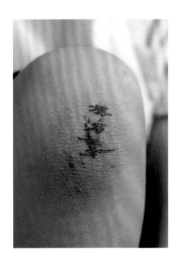

*We are only able to heal from cuts and scrapes because our body can make new skin cells to replace those that were damaged.*

## MITOSIS

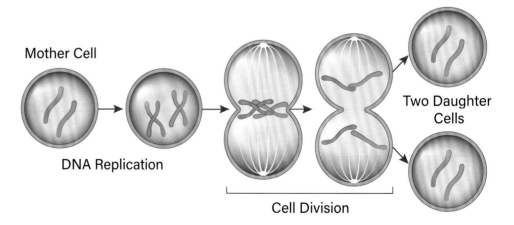

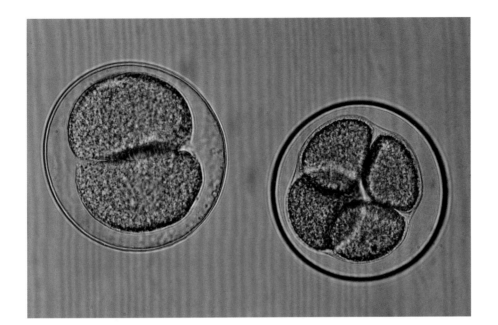

*These images show the process of cell division occurring in a sea urchin embryo.*

can make the cookies too. It would be silly for you to keep half of the recipe and give your friend the other half. Then neither one of you could make the cookies because you would only have half of the recipe. Instead, you would make a *complete copy* of the recipe so that both of you have the full recipe. This is the same with the DNA in the cell; both cells need a full copy of the DNA. After the DNA is copied, they wrap tightly around proteins to form chromosomes in the cell. This will make it easier to separate the two copies of DNA.

During cell division, the copies of chromosomes are pulled to separate sides of the cell. Remember that in eukaryotic cells, the chromosomes are stored in a membrane-bound nucleus. This means that in order to separate the chromosomes, the nucleus breaks down during mitosis. Now the copies of chromosomes can be pulled to opposite ends of the cell. One set of chromosomes moves to one side of the cell, and the other set of chromosomes moves to the other side of the cell. A new nucleus forms around each set of chromosomes. Finally, the cytoplasm of the cell starts to pinch off in the middle, resulting in two new cells. The cells are exactly the same. They have the same

## How Fast Do Cells Divide?

Most human cells divide once every day, but some cells divide faster than others. For example, your skin cells take about one hour to divide. Nerve cells, on the other hand, don't divide. They form when a baby is growing in its mother's belly and those nerve cells stay with you for your entire life.

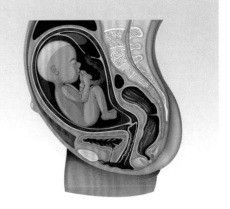

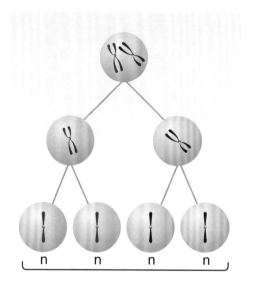

*The cell division process of meiosis is unique. You can see how it divides twice, but in the second replication/division it does so with only half the number of chromosomes. This special kind of cell division helps certain plant and animal cells to reproduce.*

genetic information and look identical to each other. These are called **daughter cells**. Once cell division is finished, the new daughter cells grow and perform their special jobs.

## MEIOSIS: A SPECIAL KIND OF CELL DIVISION

Now that you know a little about mitosis, let's look at a special kind of cell division called **meiosis** (pronounced *my-ō-sis*). During meiosis, a special group of cells will divide two times in a row to make four cells with half the number of chromosomes. So if a cell starts with ten chromosomes, after meiosis, there will be four cells with five chromosomes. This may seem kind of strange. Why would a cell want to have half the number of chromosomes? The answer is that meiosis only occurs in some very special cells in plants and animals that are used for them to reproduce.

Remember when you read about pollination in our *Plants* book? For plants to reproduce, pollen from one plant is transferred to another plant. This allows the plants to reproduce and make new seeds. Let's look at the example of a tomato plant.

Tomatoes have twenty-four chromosomes in their cells. If a tomato plant produces a pollen grain with twenty-four chromosomes, and this joins with a cell from a second tomato plant that also has twenty-four chromosomes, that makes a new cell with forty-eight chromosomes. This presents a problem, since tomato cells are only supposed to have twenty-four chromosomes. To get the right number of chromosomes in the new tomato plant, the pollen needs to start with half the number of chromosomes. The special process of meiosis takes the cell with twenty-four chromosomes and divides it to make daughter cells with twelve chromosomes. Now, a pollen grain with twelve chromosomes joins with a cell from the second tomato plant that also has twelve chromosomes. This results in a new cell with twenty-four chromosomes, which is exactly what is needed. The new cell will grow and divide through mitosis to become a new tomato plant, and the process begins again.

To make it easier to understand, imagine two boxes of crayons, both with twenty-four crayons. There is an empty box you have to fill beside these two, along with a crayon-cloning machine (okay this analogy really requires that you use your imagination!). If someone told you that you had to fill the new box with the contents of the other two, you couldn't clone *all* the crayons and fill it up, because it would be too many (forty-eight). Instead, you could clone twelve from each box and fill it to the right capacity.

## CONTROLLING CELL DIVISION

Cells do not divide all the time. They only divide when they need to. God designed our cells so that they know when it is time to divide, like for growth.

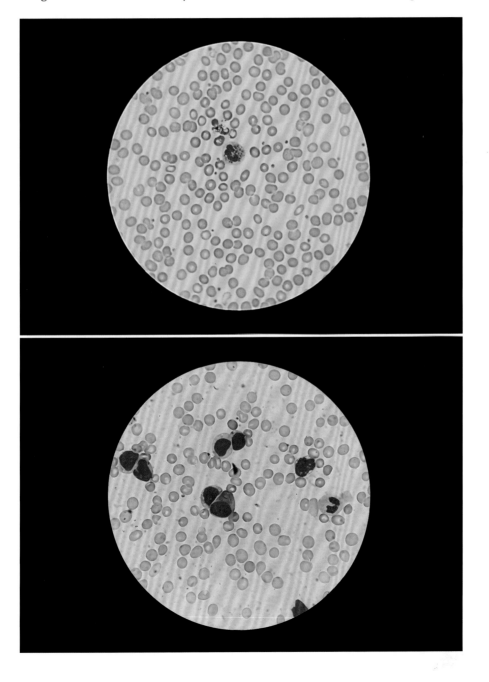

*These microscopic images show a close-up of red and white blood cells. The red blood cells are in pink, while the white blood cells are the larger purple cells.*

## Saint Peregrine, Pray for Us!

Despite the fact that we can sometimes get sick, and people are diagnosed with diseases like cancer, we should be thankful that God gives us strong bodies that can fight off infections. The human body has an immune system that can strengthen itself over time and learn to fight off sickness each time we are exposed to certain germs.

God also blesses us with His love and care during times of sickness. One way He blesses us is by giving us the saints, holy men and women we can strive to imitate and ask for their intercession before God. Saint Peregrine is the patron saint for those affected with cancer. During his life, he was diagnosed with a cancerous growth on his right foot. Just before surgery, his foot was miraculously cured. People already knew Peregrine to be a holy man, and this miracle only confirmed it. He would later bring about miraculous cures of other people, sometimes by simply whispering "Jesus" into their ears.

But our cells will not divide if they are damaged or if there is something wrong with them. If something is wrong, the cell will try to fix the problem so it does not pass the problem on to the new daughter cells. For example, cells sometimes make a mistake when copying the DNA. When this happens, the cell will try to fix the mistake in the DNA. If the problem cannot be fixed, then the cell will self-destruct. Why would a cell destroy itself? This happens so that the damaged cell does not make more damaged cells (it doesn't copy itself and thus copy the mistake). Isn't that fascinating! Let's look at two examples of what happens when things go wrong with cell division.

If mistakes in the cell cannot be fixed, then it could cause a cell to divide too much. This can result in too many cells being made and might lead to **cancer**. Cancer occurs when there is uncontrolled cell growth. Unfortunately, people can die from cancer. But doctors can do a lot of things to help treat cancer too. The first thing they would try to do is remove the cancer cells that should not be there. They can also use strong medicines called **chemotherapy** to kill the cancer cells. When cancer is found early (before the cells can reproduce too much), there is a good chance that the patient will be able to fight off the cancer with the help of the doctors and medicine. But this can make them feel very tired and sick while their body is working hard to get rid of the cancer.

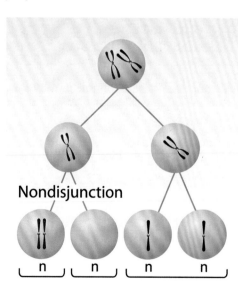

*Nondisjunction occurs when one or more pairs of chromosomes fail to separate as they should. (note the red versus the blue). This abnormality can cause health problems and harm human development.*

Another mistake that sometimes happens during cell division is that the chromosomes do not separate properly. This is called **nondisjunction**. When this happens, the daughter cells might end up with an extra chromosome, or one fewer chromosome. Usually if the cell does not have the correct number of chromosomes, it is not able to function properly and will not survive. But there are a few special cases of cells with an extra chromosome being able to grow and divide. Normal human cells have forty-six chromosomes (twenty-three pairs—23 x 2 = 46—which we refer to by their number). But because of a mistake during cell division, some people end up with forty-seven chromosomes in their cells. When there is an extra third copy of chromosome 21, this results in a condition known as **Down syndrome**. If you have ever met someone with Down syndrome, you know that they are incredibly happy, kind, and loving individuals. And most importantly, they are made in the image and likeness of God, just like you and me.

In the next chapter, we will start to look at how cells work together for a common purpose.

*Cells Fun Fact:*
*New cells that are created from existing ones are called daughter cells.*

## FOUNDATIONS REVIEW

✓ All forms of life must be able to make more of itself (reproduce) in order to survive. Cells do this through a process called cell division. When a cell divides, it must make a copy of all of its DNA. Each new cell will need a full copy of the DNA in order to function properly. The cell makes a perfect copy of the DNA in a process called DNA replication. New cells that are created from existing ones are called daughter cells.

✓ God designed our cells so that they know when it is time to divide. But our cells will not divide if they are damaged or if there is something wrong with them. If something is wrong, the cell will try to fix the problem so it does not pass the problem on to the new daughter cells. If it cannot fix the problem, the cell will self-destruct.

✓ Despite cells doing what they can to divide and reproduce properly, there can be problems. The disease known as cancer occurs when there is uncontrolled cell growth, while the condition known as Down syndrome happens when cells, which in humans normally have forty-six chromosomes, end up with an extra chromosome through faulty cell division.

# Dr. Jerome LeJeune

Dr. Jerome Lejeune was a French physician who lived from 1926 to 1994. In addition to being a doctor, Dr. Lejeune was also a research scientist and a devout Catholic. During his life, there were a group of children that he worked with known, at the time, as "mongoloids." They were given this name because of some of their unique facial features, such as a small, flattened nose and almond-shaped, slanting eyes. These children also had some delays in their growth and their ability to learn. Because these children were not well understood, they were often hidden away from the rest of the world. Oftentimes their families were even blamed for their condition. Dr. Lejeune wanted to help these children and their families by trying to learn what caused them to have this disease and to try to find a treatment for them.

In his studies, Dr. Lejeune discovered that "mongolism" results when a child is born with an extra chromosome in their cells. Today we call this condition Down syndrome. For his work in discovering the cause of Down syndrome, Dr. Lejeune was given the Kennedy Prize in 1962. Later, in 1969, he was awarded the William Allan Award, the highest award in the field of genetics.

The work of Dr. Lejeune was very important in the field of genetics. It also led to research that allows mothers to know if their babies have Down syndrome before they are born, a process called pre-natal testing. Tragically, this information is often used by mothers who choose not to give life to their babies.

Dr. Lejeune was a devout Catholic who saw each life as a precious gift from God, so the idea that people were making these decisions was very upsetting to him. He became a strong advocate for the pro-life cause, giving hundreds of talks and interviews throughout the world. Through his work, Dr. Lejeune met Pope John Paul II, who appointed him as the first president for the Pontifical Academy for Life. Dr. Lejeune only served as president for a short time (thirty-three days) before he died of lung cancer in April 1994.

In April 2012, the cause for canonization was opened for Dr. Lejeune, giving him the title "Servant of God." Then, on January 21, 2022, Pope Francis recognized the heroic virtues of Dr. Lejeune when he elevated him to the state of "Venerable."

You can read more about the amazing life and work of Dr. Jerome Lejeune in his biography, *Life is a Blessing*, written by his daughter, Clara Lejeune.

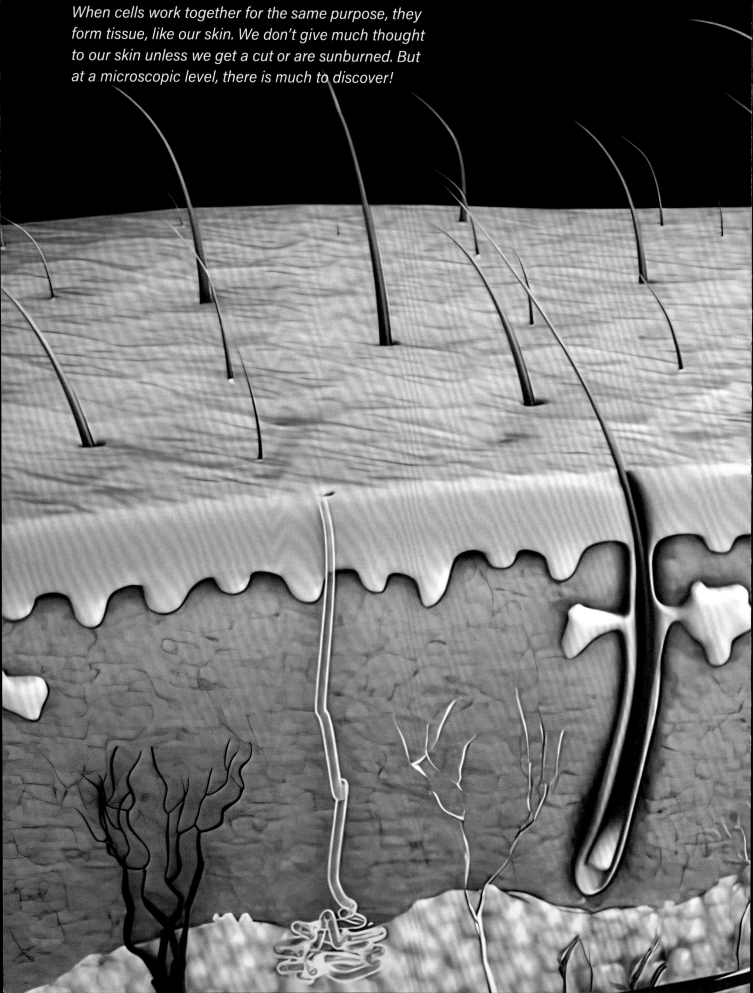

When cells work together for the same purpose, they form tissue, like our skin. We don't give much thought to our skin unless we get a cut or are sunburned. But at a microscopic level, there is much to discover!

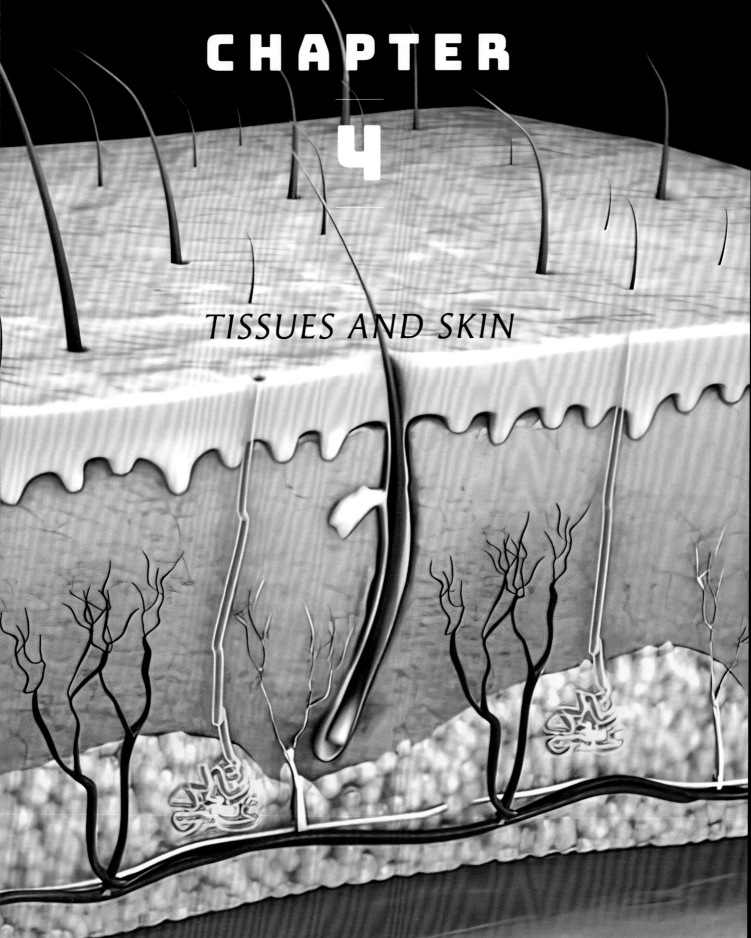
# CHAPTER 4
## TISSUES AND SKIN

**FOUR TISSUE TYPES**

Connective Tissue

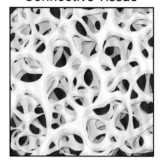

Neural Tissue

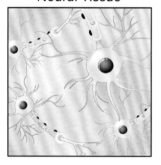

Epithelial Tissue

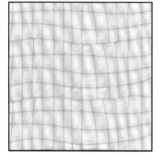

Muscle Tissue

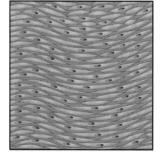

## WHAT IS A TISSUE?

In the last three chapters, we have spent a lot of time learning about cells, their structure, and how they divide. Cells are the building block of all living things, including us. In organisms, such as humans, cells must work together to form tissues and organs. In this chapter, we are going to learn how cells work together to carry out a common purpose.

To begin, imagine you have a big, warm, heavy quilt that covers your bed to keep you warm at night. The quilt is made of many smaller pieces of fabric that have been sewn together. Individually, those pieces of fabric don't do a very good job to keep you warm. But when they are sewn together into a quilt, you have a warm blanket. The pieces of fabric together take on a new role that they did not have—could not have—on their own. This is similar to how our cells work together to carry out a special job. When similar cells work together for the same purpose, then we say they form a **tissue**.

There are four different types of tissue found in our bodies. **Epithelial tissues** are tissues that cover. Our skin is an example of epithelial tissue. Skin, or flesh, covers our entire body. It also lines the inside of our stomach, blood vessels, and other organs. We will talk more about skin later in this chapter. But first, let's learn the other three types of tissue. The second type of tissue is **connective tissue**. Connective tissue connects, binds, or supports. In other words, connective tissue holds other types of tissue together. An example of connective tissue is bone. We will learn more about bone in chapter 5. The third type of tissue is **muscular tissue**. In chapter 6, you will read more about muscles. The last type of tissue is **nervous tissue**. This tissue is formed from many special cells coming together to form nerves. Your brain is made of nervous tissue. We will talk about nervous tissue in chapter 7. These four different types of tissues are found all throughout our bodies. Oftentimes different tissues work together to form organs such as our heart, brain, stomach, and skin.

## SKIN

Did you know that your skin is the largest organ in your body? You probably don't even think of skin as an organ, but it is. In a nine-year-old, your skin would take up about as much space as a cardboard table. Skin is made up of two main layers and can be up to four millimeters thick. The outer layer of skin is called the **epidermis**. It is made of many layers of cells. The layers of cells at the top, the ones that you can touch, are all "dead" cells. The dead cells fall off and have to be replaced by new cells that grow up in their place. When you have dry skin in the winter and it flakes away, dead cells are falling away, though you lose cells without even seeing or noticing too. Some estimates say that a person loses thirty thousand dead skin cells every day! But don't worry, those cells that fall off are replaced. The very bottom layers of cells in the epidermis are constantly dividing to replace the cells that fell off. In fact, you get a "new" suit of skin about once a month. Our skin cells contain a lot of a special protein called **keratin** that protects it and keeps it from tearing (though of course sometimes we do get cuts and scrapes).

The bottom layer of your skin is called the **dermis**. The dermis is much thicker than the epidermis. It contains special cells called **nerves**. It also contains blood vessels, sweat glands, and hair follicles. The nerves in your skin let you feel things like the coldness of an ice cube, the heat of a hot pan, or the prick of a needle. By being able to feel, your skin can protect you from hurting yourself. Imagine what would happen if you could not feel pain. If you could not feel the burn of putting your hand down on a hot stove, you would leave your hand there and the skin would roast off. Being able to feel the burn is a valuable form of protection, or self-defense. Being able to feel is also very important because it helps us to be able to hold a pencil or other objects correctly.

Meanwhile, the blood vessels in our skin help keep our body temperature steady. When we get too warm, more blood runs through the blood vessels in our skin. This releases heat from our bodies and helps us cool off. Another way our skin helps us to cool off is that it contains tiny little sacs called **glands**. These glands produce sweat when we get hot. The sweat evaporates and helps cool us off.

Though we said there are two *main* layers to our skin, there is a third layer at the bottom called the **hypodermis**, below the dermis. This is made mainly of fat tissue, which helps keep us warm.

## HAIR AND NAILS

As you probably know, hair and nails grow out of our skin. They actually contain the same keratin protein as our skin, but in our skin, the keratin is soft and flexible, while in our nails, the keratin is held together tighter to make it much stronger. Nails help to protect the tips of our fingers and toes from getting damaged. They can also be useful for scratching an itch (or peeling a sticker off

*In this diagram, you can see the different layers of our skin. Note how the hair follicles are rooted all the way down in the dermis near where your nerves are. This is why it hurts to have the hairs on your body pulled out!*

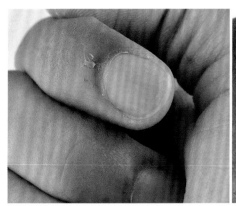
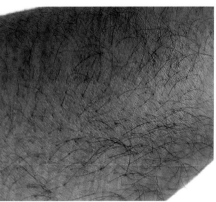

## Why Does My Skin Get Wrinkly in the Tub?

When you sit in a bath for a long time, the dead cells that make up most of the epidermis absorb water. When cells absorb water, they get bigger. This causes your skin to look bumpy or wrinkled. You notice the wrinkles more in areas where your skin is thicker, such as the skin on your hands and feet.

The wrinkles you get in the tub are different than the fingerprints you find on the tips of your fingers. These are caused by ridges in the dermis. They formed when you were still growing in your mother's womb. Everyone's fingerprints are different. They are unique to just you, because God only created one you!

a toy!). Meanwhile, hair grows out of structures in our dermis called **hair follicles**. This is similar to how a tulip grows up out of a bulb. Some of our hair is long and coarse, like the hair on our head. But most of the hair on our skin is short and fine. Look closely on your arms or legs; you can see the little hairs growing out of your skin. When you get cold, those hairs stand up on your arms to try and help you get warm.

People have different colors of hair and skin. The color comes from a special brown pigment called **melanin** that is found deep in the epidermis. Special cells in the bottom layers of the epidermis make melanin. People have different amounts of melanin in their skin. Someone with a lot of melanin will have darker skin. If someone has very little melanin, their skin has a pinker appearance because the color of the blood shows through their skin. Freckles form when there is a lot of melanin that collects in one place in the skin. These are easier to see on people with paler skin. You may have noticed too that your skin turns darker the more you stay in the sun; this is because your skin increases its production of melanin to protect it from the harmful rays of the sun.

*Due to the varying amounts of melanin we have in our skin, we find people with different skin tones, and even our own skin can slightly change color if we spend time in the sun because it produces more melanin as a protection against the sun's rays. It's important to wear sunblock when we go to the pool or the beach—too much UV radiation can damage the DNA in our skin cells, which can cause cells to grow out of control and lead to skin cancer.*

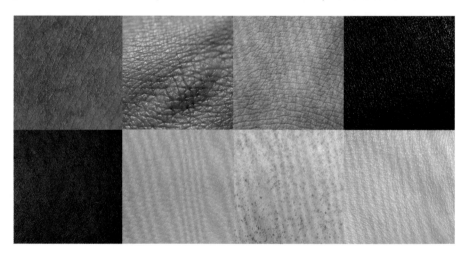

## THE FUNCTIONS OF SKIN

Skin is very important for our health and survival. One of its most important jobs is to help keep things out of our body that might want to get inside. These things could be bacteria, germs, or harmful chemicals. If we get cuts or breaks in our skin, this provides openings for bacteria to get into our bodies that can cause infections or make us sick. You might compare it to a breach in a castle wall that lets invaders in.

Skin also helps to protect us from other things in our environment, such as toxins or UV radiation from the sun. However, we have to be careful because too much sun could damage our skin. This is what happens when we get a

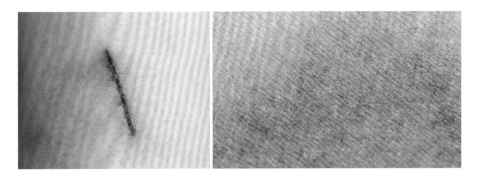

*Our skin can be damaged by cuts or sunburn, but through cell regeneration, it can heal itself in just a matter of days! In the case of sunburn, you can actually see the dead cells flaking away as your skin peels off.*

sunburn. The sun's strong rays damage the skin and cause it to turn red and become painful. That is why it is important to protect our skin from the sun by wearing hats, lightweight clothing, or using sunscreen.

Our skin also helps to keep water inside our bodies, which keeps us hydrated (human beings can become very sick or even die if we lose too much water and become dehydrated). Our bodies contain a lot of water, which is important for our cells, tissues, and organs to function properly.

Now that we have covered our skin, let's go deeper into our bodies and examine our bones and skeletal system.

## FOUNDATIONS REVIEW

✓ When similar cells work together for the same purpose, then we say they form a tissue. There are four different types of tissue found in our bodies. Epithelial tissues are tissues that cover, like our skin. Connective tissue connects, binds, or supports. The third type of tissue is muscular tissue, and the last type of tissue is nervous tissue. This tissue is formed from many special cells coming together to form nerves.

✓ Skin is the largest organ in our bodies. It is made up of two main layers and can be up to four millimeters thick. The outer layer of skin is called the epidermis. The cells that make up this layer can "die" and fall off, but they are replaced by new cells. The bottom layer of your skin is called the dermis, which is much thicker than the epidermis. It contains nerves, as well as blood vessels, sweat glands, and hair follicles. Finally, below the dermis is another layer called the hypodermis. This is made mainly of fat tissue, which helps keep us warm.

✓ Skin is very important for our health and survival, carrying out several important functions: (1) It keeps things out of our body that might want to get inside. These things could be bacteria, germs, or harmful chemicals. (2) Skin also helps to protect us from other things in our environment, such as toxins or UV radiation from the sun. (3) Finally, our skin also helps to keep water inside our bodies, which keeps us hydrated.

# Saint Damien and the Lepers

*"It is at the foot of the altar that we find the strength we need in our isolation."*

–Saint Damien of Molokai

In the Bible, we read of several instances where Jesus encounters and heals a leper, or a group of lepers. **Leprosy** is a disease caused by bacteria attacking cells in the skin and nerves. People who have leprosy develop flaky patches on their skin. Because the nerves are damaged, they also lose sensation in their skin, fingers, and toes. This means they are unable to feel and may cut, burn, or damage their skin without knowing it. Over time, the body may even start to reabsorb the fingers or toes because of the loss of nerves. This makes it look like their toes or fingers have fallen off. People with leprosy suffer a great deal.

Leprosy is a **contagious disease**. This means that the bacteria can be spread from one person to another. Thankfully, it is not easily spread, and we have medicines called antibiotics to treat it. As a result, even though it may have been more common in the time of Christ, leprosy is very rare in our modern world.

During Jesus's time, people did not know what caused leprosy, and they did not have a way to treat it. Lepers had ugly looking skin and disfigured fingers, toes, and faces. Other people did not want to be around the lepers because they were afraid they might also develop leprosy. Jewish laws found in the Old Testament instructed lepers to live alone, outside of the camp. They were considered "unclean" and could not participate in the Jewish life or rituals. When Jesus healed the lepers, He not only healed their physical sickness but He also allowed them to be "clean" again. Now they could live with other people and attend the Jewish religious ceremonies.

Saint Damien of Molokai is the patron saint of those with leprosy. Saint Damien was born in 1840 into a farming family in Belgium. He discerned a

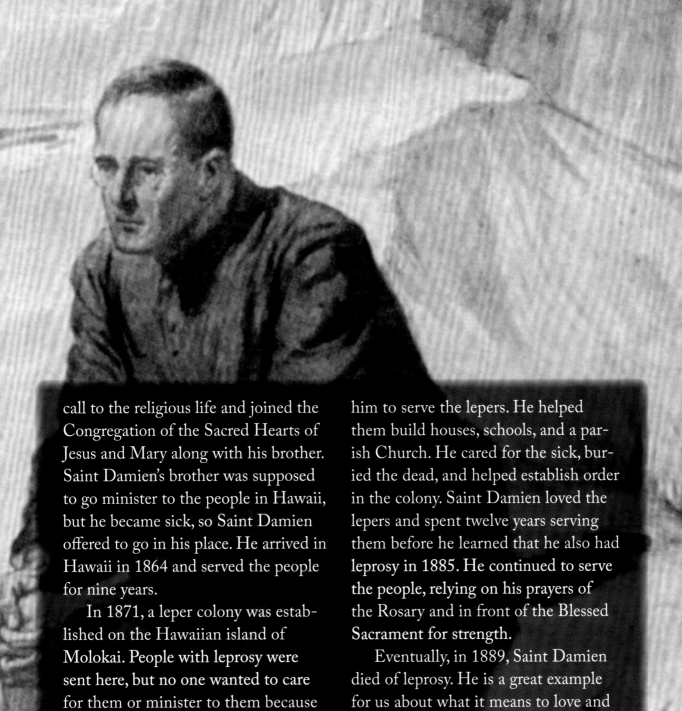

call to the religious life and joined the Congregation of the Sacred Hearts of Jesus and Mary along with his brother. Saint Damien's brother was supposed to go minister to the people in Hawaii, but he became sick, so Saint Damien offered to go in his place. He arrived in Hawaii in 1864 and served the people for nine years.

In 1871, a leper colony was established on the Hawaiian island of Molokai. People with leprosy were sent here, but no one wanted to care for them or minister to them because they were afraid they would get leprosy. Saint Damien discerned God calling him to serve the lepers. He helped them build houses, schools, and a parish Church. He cared for the sick, buried the dead, and helped establish order in the colony. Saint Damien loved the lepers and spent twelve years serving them before he learned that he also had leprosy in 1885. He continued to serve the people, relying on his prayers of the Rosary and in front of the Blessed Sacrament for strength.

Eventually, in 1889, Saint Damien died of leprosy. He is a great example for us about what it means to love and serve others, especially the neediest. Saint Damien, pray for us!

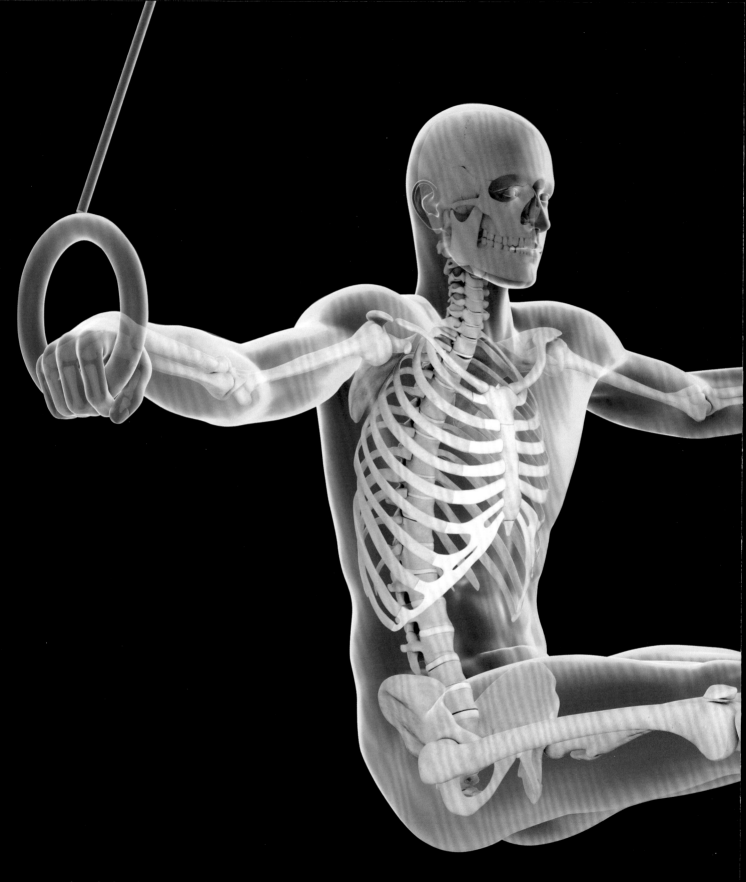

*The skeletal system gives our bodies shape and support, works with our muscles to help us move around, protects our vital organs, stores important resources like calcium and fat, and is where our blood is made.*

# CHAPTER 5

## THE SKELETAL SYSTEM

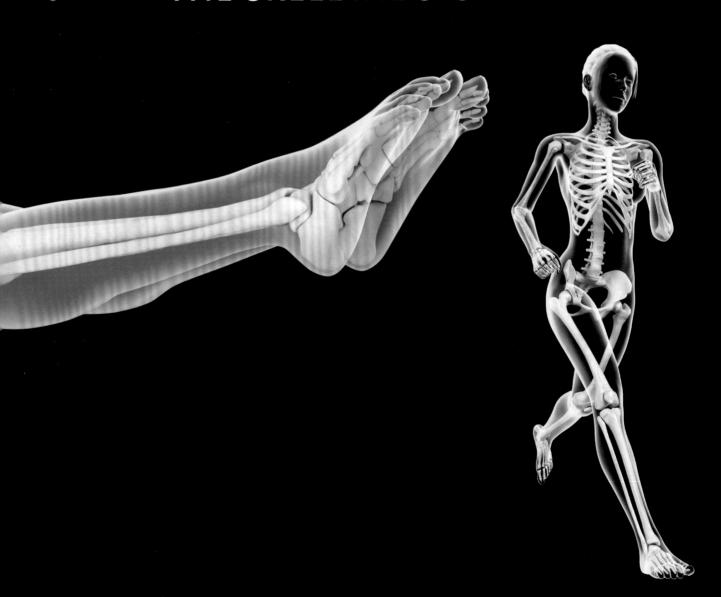

*Skeletal System Fun Fact:*
*A typical adult has over two hundred bones in his or her body!*

## FUNCTIONS OF THE SKELETAL SYSTEM

Now that we have covered the structure and function of our skin, it is time for us to begin our exploration of the systems *inside* our body. We will start by looking at the framework of the human body, known as the **skeletal system**. The skeletal system gives our entire body its structure and shape. Without the bones that make up the skeletal system, we would be a blob of soft tissue creeping along the ground.

Our skeletal system plays a very important role in protecting internal organs such as our heart and brain, which we will learn more about later on in our book. Still more, our bones work together with our muscles to help us move around. Finally, our bones help to store important resources such as calcium and fat, and they are also where our blood is made. But these are very broad functions of the skeletal system; let's get more specific and go into greater detail so we can understand more about this vitally important part of our bodies.

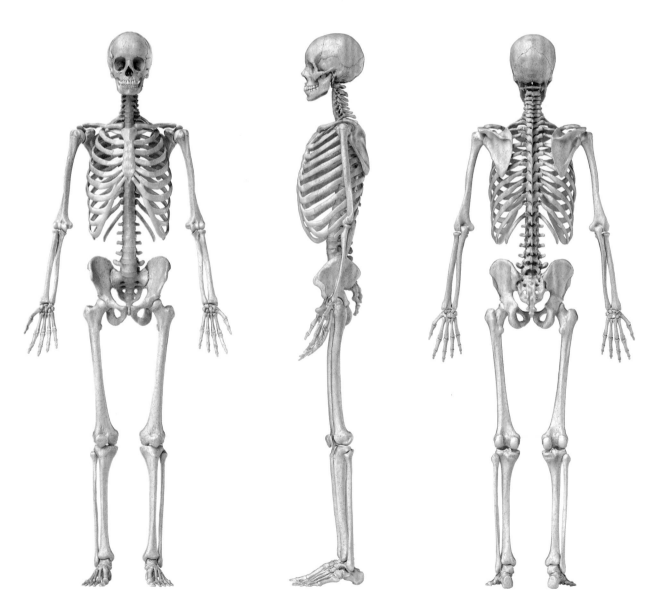

## WHAT IS BONE?

What image comes to your mind when you think of a bone? Perhaps you think of the type of bone your dog might bury in the yard. Or maybe you think of the chicken bone that is left over after eating a drumstick of fried chicken. These are bones, to be sure, but they don't give us a very clear picture of the bones living and growing inside us, and those are the ones we care about here.

**Bones** are living tissues made of thousands of bone cells called **osteocytes** (*cyte* is a common suffix that refers to cells). The osteocytes secrete and build a matrix, or frame, of hard material made of calcium and other minerals. The minerals are laid down on a framework of protein fibers called **collagen**. The collagen fibers are elastic and bendy (like a rubber band), while the calcium and minerals are hard. These two materials work together to make bones that are rigid and strong but can also bend just enough so that they do not easily break or crumble like chalk.

If you were to look at bone under a microscope, you would see that the minerals are laid down in a pattern of circles. In fact, it looks a lot like a tree stump that is left behind after cutting down a tree. Each of these sets of circles is called an **osteon**. The small dark slits you see between the rings are little "windows" called **lacunae** that house the osteocytes (the bone cells). In the middle of the osteon is the central canal. This contains blood vessels and nerves. The blood vessels bring food and oxygen to your bones so they can continue to grow.

### A Boney Kind of Activity!

The next time your mom or dad cut up some chicken for dinner, ask them to help you with this fun activity. Gather some of the chicken bones and place one in a cup of vinegar overnight. The vinegar is an acid that will eat away at and remove the calcium salts in the bone. After a while, the bone will soften such that you can bend it without it snapping or breaking (but it won't be very sturdy now). This is because the collagen fibers are very flexible. Next, take another bone and boil it in hot water for a few hours. The boiling removes the collagen, leaving the calcium salts behind. This bone will become very brittle and will snap in half if you try to bend it. These experiments show how the collagen fibers and calcium salts work together to make our bones what they are. Taking away one or the other results in bones that do not function as well.

*This activity is repeated in the Cells & Systems Workbook.*

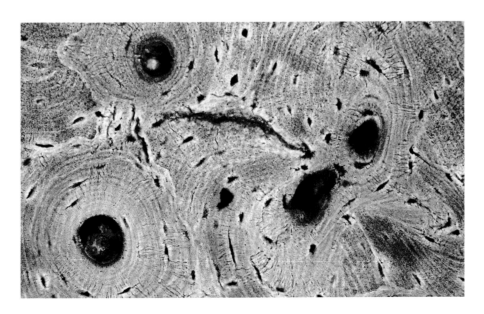

*Under a microscope, bone tissue presents itself in concentric circles, similar to the rings seen on top of a tree stump.*

## THE FORMATION AND GROWTH OF BONES

You may be wondering how bones grow. After all, you are bigger than you were last year, and will be bigger still next year, right? Remember that bone is a *living* tissue made of cells. As those cells divide, they produce more bone cells, which causes your bones to grow. But perhaps we should begin by looking at how our bones form in the first place.

In order to make new bone, the osteocytes need a frame on which to lay down the calcium and other minerals needed to build bone. So before the bones form, we first see skeletal structures made of cartilage. **Cartilage** is a special type of connective tissue made of the strong protein collagen. This makes cartilage very flexible and bendy. You have cartilage in your ears and in the tip of your nose. Take a moment and bend your ear or nose and you will see what cartilage feels like. Do you see how these body parts are softer and more bendable than, say, your wrist or fingers? That is the difference between cartilage and bone.

From the time God creates a new life in the form of a baby inside his mother, it takes about two months for a skeleton of cartilage to form. Once the cartilage skeleton is in place, osteocytes begin to lay down the calcium and minerals to make up spongy bone. **Spongy bone** consists of networks of bone tissue with a lot of small spaces in between. We call this "spongy bone" because it looks kind of like a sponge. The osteocytes start putting down

*During growth, new bone cells continue to form at the growth plate, lengthening the bones. But as we start to reach our full height, the thin layer of cartilage is replaced by bone. At this point, we say that the growth plates have "closed" and the bones stop growing longer. Otherwise we would grow forever!*

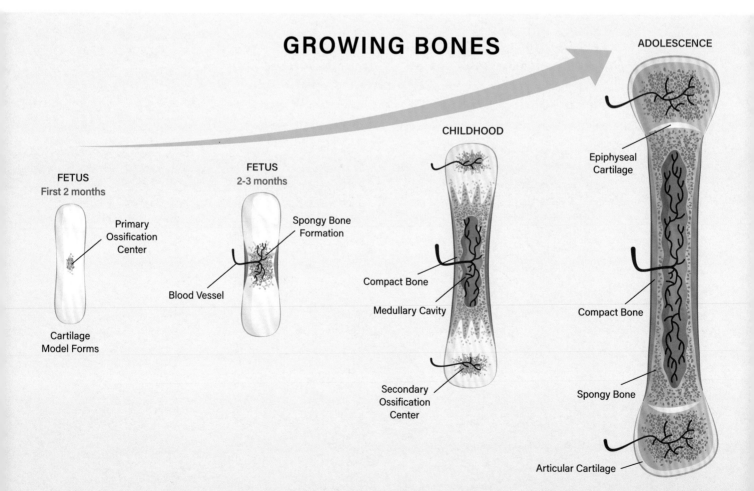

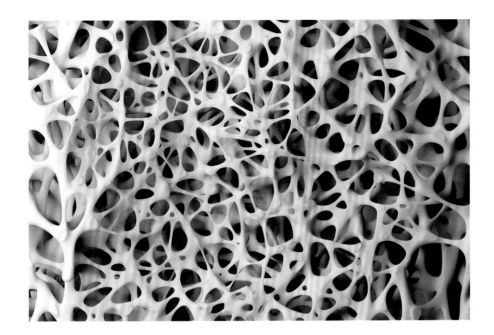

*You can see where spongy bone gets its name—when we look at it under a microscope, it looks very much like a sponge!*

spongy bone in the middle of the bones, and then it continues to build it up and down the length of the bone. As the bone continues to form, the cartilage is removed. The outer part of the bone is made of thick layers of compact bone, like we saw in the microscope image in this section. By the time the baby is born, the baby's skeleton is composed mostly of bony tissue. And as the baby grows into a toddler and then a young child and eventually into an adolescent, the cartilage will continue to be replaced by bony tissue.

## Building Up Strong Bones: Diet and Exercise

Building strong bones is very important. You can help your body make strong bones by having a diet of a lot of vegetables and food with calcium (a diet here does not mean trying to lose weight, like you might hear adults say, but rather all the various things you eat). A good source of calcium is milk and other dairy products. It is also important for you to be active. When you use your bones and muscles, this puts stress on your bones. The osteocytes respond by laying down more bone tissue. This makes your bone stronger.

Your bones continue to grow bigger, which is what allows you to grow bigger and taller. Between the middle region of long bones, called the shaft, and the "knobs" at the ends, is a thin layer of cartilage which we sometimes call the **growth plate**. While you are growing, new bone cells continue to form at the growth plate. This causes the bones to lengthen. As a person starts to reach his full height, the thin layer of cartilage is replaced by bone. At this point, we say that the growth plates have "closed" and the bones stop growing longer. This is why we eventually stop growing; otherwise, we would all be giants! In addition to getting longer (or "taller"), bones can also grow thicker and wider. Osteocytes continue to replace old bone tissue to make it stronger.

## SHAPES AND SIZES OF BONES

Bones come in many different shapes and sizes. Long bones are hollow cylinders with knobs on their ends (just think of the shape of the bones you give your dog). Inside the **long bones** there is a substance called yellow marrow which stores fat. Meanwhile, the knobs of the long bones look a bit different inside. Instead of a hollow space, they contain spongy bone. This is where

*When we picture a bone in our head, we are probably imagining what we call "long bones." But there are many other types of bones, each serving their own unique purpose in accordance with their shape and size.*

# BONE CLASSIFICATION BY SHAPE

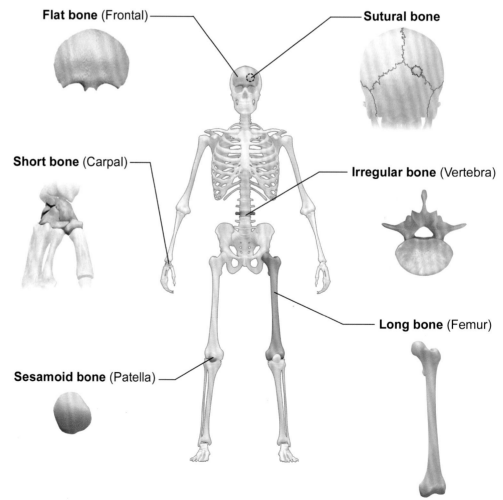

blood is made. Some examples of long bones are the bones in your legs, arms, fingers, and toes.

Other types of bones are the short bones. These are about as wide as they are long. The bones in your wrists and ankles are examples of **short bones**. The bones that make up the skull in your head and your ribs are **flat bones**. They are shaped more like a plate. Then there are some bones which have odd shapes, like the vertebrae that make up your backbone. These are called **irregular bones**.

## THE SKELETON: CONNECTING BONES

Now that we have learned about individual bones, let's see what they look like when they are all put together.

Did you know an adult has 206 different bones in his body? To show the range, the largest is the femur in the leg, which is just under twenty inches on average, while the smallest bones are found inside our ears at only a few millimeters.

Newborn babies have somewhere between 270 and 300 bones when they are born. This is because their skeleton is not fully formed. As the baby grows, some of his bones start to join and fuse together to become a single bone, which explains why a baby would have more bones than an adult. One example of this can be observed in the head of a newborn baby. The human skull is made of several bones that fuse or join together. In a newborn baby, those bones are separate. The spaces between the bones are filled with cartilage. The most prominent of these spaces is the "soft spot," also called the **fontanelle**, on top of the baby's head. If you were to gently rub your fingers over a baby's head, you may be able to feel the spot where the bone has not yet grown together. This makes it much easier for the baby to be born because his head can change shape a little bit.

The skeletal system is divided into two parts, the axial skeletal system and the appendicular skeletal system. The **axial skeletal system** includes the skull and vertebral column. As we just stated, the skull is made of many individual bones that are fused together. These help to protect your brain. The skull also includes the mandible, or jaw bone, that allows you to open and close your mouth. This is what makes up your chin. The vertebral column consists of thirty-three bones, with the top twenty-four vertebrae making up your neck and back. Have someone bend over to touch his toes and you can see the vertebrae that make up his backbone, or spine. The other seven bones are fused together to make up the sacrum and coccyx. We usually refer to the coccyx as the "tailbone."

The **appendicular skeleton system** includes all of the other bones. The figure on the next page shows some of the major bones of your body. You can feel some of these bones. For example, the clavicle, also known as the collar bone, is on the front of your shoulder where you find the collar of your shirt. If you put your hand behind your back, you may be able to reach up and feel

*Remember:*
*One of the most important functions of the skeletal system is that they protect our vital organs. So, for example, our skull protects our brain, and our ribs protect our heart. You can almost think of it like the shell that protects the turtle!*

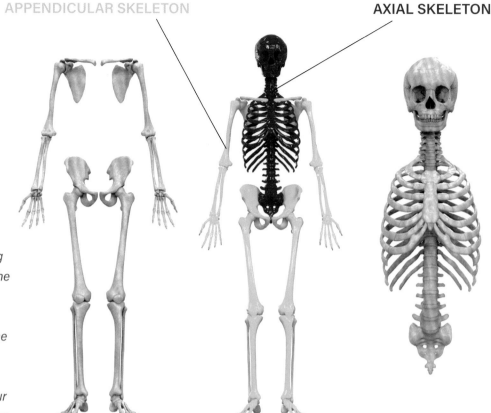

**APPENDICULAR SKELETON**  **AXIAL SKELETON**

*An axle in a car is the long rod that passes through the center of the wheels and connects them. Similarly, an axis is an imaginary line around which an object rotates. Thus, it makes sinse for our vertebrae (our spine) to be included in the axial skeletal system, as it is the long, narrow set of bones that runs through the center of our bodies.*

your shoulder blade. This is part of the bone called the scapula. It connects to the humerus bone, which is the main bone in the upper arm. Your elbow is the end of the bone called the ulna, which is one of two bones in your forearm. The other bone is called the radius. You can feel the end of the radius on the thumb side of your wrist. The pelvis, or hip bone, helps to connect your vertebral column to the femur, the bone in your upper leg. You can also easily feel the patella, or knee cap. Moving down, the shin bone is referred to as the tibia. And your ankles are the ends of the tibia as well as another bone called the fibula.

Remembering all the names of these bones is not important right now. But just touching on them briefly shows how complicated and intricate God made our skeletal system. Each of these bones serves a very important function. If we were to lose one or break one, even the smallest of them, it would make our lives much more difficult.

## JOINTS: MOVING BONES

Bones wouldn't do us much good if they didn't fit together. This is where joints come in. A **joint** is where two or more bones come together. You might have tinker toys or something like them where long stick-like pieces can be put together to form bigger shapes; except that these pieces cannot actually come together unless you have the connecting pieces as well, usually a ball or square with holes in it to receive the sticks. These connecting pieces are sort of like the joints in our bodies that connect our bones.

Some joints fit more tightly. For example, the femur has a rounded end that fits into a bowl-shaped hole in the pelvis bone. This makes it a very stable joint that can hold up our body weight. Other joints, like our shoulder, are more flexible. This makes it so we can swing on the monkey bars, swim, and throw a ball. However, the shoulder joint is not as good at holding large amounts of weight. We call these ball-in-socket joints. The knee and elbow are hinge joints. They bend and straighten similar to how a door might open and close. Joints are held together by strong bands of connective tissue called **ligaments**. The ligaments help to keep the bones from slipping out of place. Muscles also play an important role in helping to stabilize joints.

Speaking of muscles, it is to them that we will turn in our next chapter.

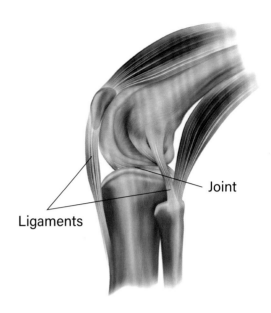

## FOUNDATIONS REVIEW

✓ Our skeletal system plays a very important role in protecting internal organs such as our heart and brain. Still more, our bones work together with our muscles to help us move around. Finally, our bones help to store important resources such as calcium and fat, and they are also where our blood is made.

✓ Bones are living tissues made of thousands of bone cells called osteocytes. They are made of a mix of collagen fibers and calcium salts (among other minerals), which give them both their sturdiness and their ability to bend just enough to keep them from breaking easily or crumbling. Since bone is a *living* tissue made of cells, the cells can divide and produce more bone cells, which form at what is called the growth plate. This explains why our bones grow (though our growth plates eventually close, which explains why we stop growing at some point).

✓ Bones come in many different shapes and sizes, including long bones (legs, arms), short bones (wrists, ankles), flat bones (skull, ribs), and irregular bones (vertebrae/backbone). Overall, there are over two hundred bones in an adult human body, all of which connect with joints. The skeletal system in total is divided into two parts (two different sets/kinds of bones): the axial skeletal system and the appendicular skeletal system.

# Saints, Prophets, ... and Bones

Bones and skeletons are often associated with Halloween and portrayed as being scary. And yet it is not uncommon to see saints pictured with a skull or near skeletons. Some common examples are St. Jerome and St. Francis. The presence of the skull reminded these saints that death could come at any time, that their earthly life was only temporary. As a result, they pursued something that has eternal significance: a life of holiness.

Every year on Ash Wednesday we are reminded that, at some point, we will all die. This happens when the priest traces a cross on our forehead with ashes and says the words, "From dust you were made, and to dust you shall return." This ritual reminds us that just as God created Adam and all mankind from the Earth, our bodies will not last forever and will return to the dust of the Earth when we die. For this reason, we should live our lives to the best of our abilities so that we can attain life with God in heaven.

There is a story in the Old Testament about a prophet named Ezekiel. During his time, the Israelite people were captives in Babylon for many years. They had lost hope and in some sense were dead to the world. We read that the Spirit of God brought Ezekiel to a valley full of dead, dry bones.

He tells us, "Then [God] said to me, 'Prophesy to these bones, and say to them: O dry bones, hear the word of the LORD. Thus says the Lord GOD to these bones: I will cause breath to enter you, and you shall live. I will lay sinews on you, and will cause flesh to come upon you, and cover you with skin, and put breath in you, and you shall live; and you shall know that I am the LORD'" (Ez 37:4–6).

Ezekiel obeyed, and the bones came together and had skin, but they were not alive until God told Ezekiel to ask for the "breath to come into them". Again, Ezekiel obeyed, and suddenly they were alive.

This breath of life should remind us of the breath that God breathed into Adam in Genesis when He formed him out of the dust. Biblical scholars also suggest that Ezekiel's story can teach us about Pentecost, when God sent the Holy Ghost to the Apostles. Before this, the Chosen People were like dry bones walking around, but when they received the breath of God—the Holy Ghost—they became spiritually alive and were able to enter into the promised land of heaven, just like the Israelites could return to their promised land in the Old Testament.

Next time you encounter some old dry bones, or a skeleton, let it remind you that we have been given life through Christ and through the Holy Ghost. And through this life-giving breath, we have the supernatural hope of one day returning to our home with God in heaven.

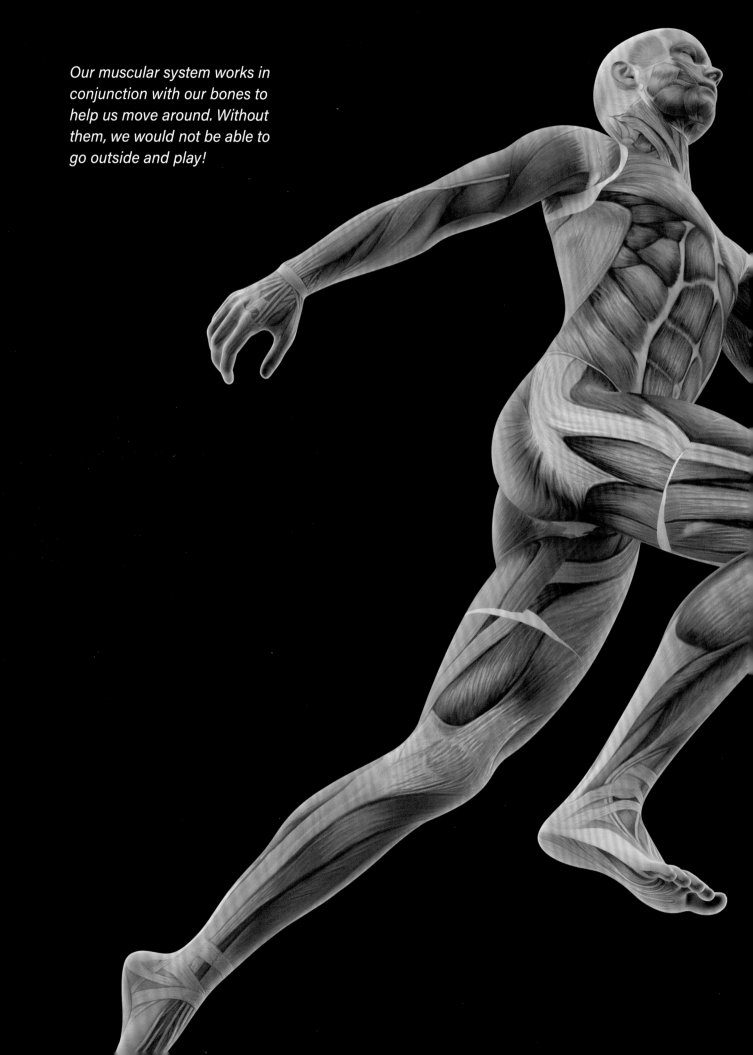

*Our muscular system works in conjunction with our bones to help us move around. Without them, we would not be able to go outside and play!*

# CHAPTER 6

## THE MUSCULAR SYSTEM

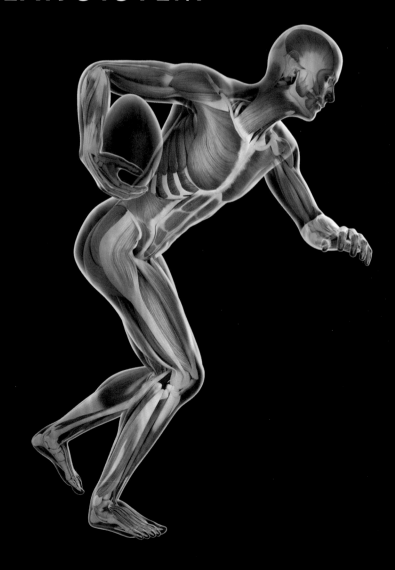

# A BODY FULL OF MUSCLES

Did you know the human body has over six hundred different muscles? The muscles that make up our **muscular system** are important in stabilizing joints, helping us have good posture, and keeping a steady body temperature. But the main function of muscles is to help us move, and this is usually what we think of when we think of muscles—those that work with our skeletal system to help us move. These are called **skeletal muscles**. Skeletal muscles are **voluntary muscles**, meaning that we can think about moving them, and then they contract, or shorten (your hand "contracts" when you make a fist and squeeze). The shortening of the muscles causes the bones to move, and when our bones move, we move. We will learn more about how that happens later in this chapter. For now, let's talk about other types of muscles in our body.

VOLUNTARY MUSCLE

Contrasted to voluntary muscles are, you guessed it, **involuntary muscles**. There are two types of muscles that are involuntary: cardiac muscle and smooth muscle. **Cardiac muscle** is the tissue that makes up the heart. The muscle in the heart contracts, causing the heart to beat. **Smooth muscle** is found inside the walls of many of our digestive organs, such as our stomach and intestines. These muscles contract to help mix up the food in our stomach and push it through our digestive system. Involuntary muscles contract when they are needed and without us thinking about it. Thank goodness we don't have to remember to have our heart beat, or we might forget!

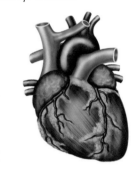

INVOLUNTARY MUSCLE

Now that we understand what sorts of muscles we have and their primary types, we should learn how muscles contract. To do that, we must first understand more about their structural makeup.

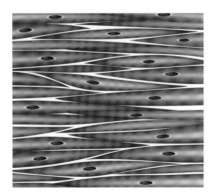

Smooth Muscle | Cardiac Muscle | Skeletal Muscle

## MAKEUP OF A MUSCLE

To better understand the makeup of our muscles, we will focus on the structure of skeletal muscles. A skeletal muscle, like the bicep in your upper arm, is a collection of hundreds of muscle fibers. These are arranged into bundles, or **fascicles**. There are several bundles of muscle fibers in a single muscle. Wrapped around the bundle of muscle fibers is a layer of connective tissue that holds them all together.

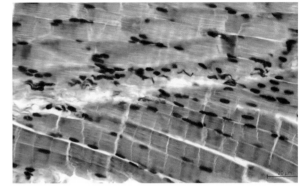

If you were to look at skeletal muscle under a microscope, you would notice that the muscle fibers are arranged in straight or parallel lines. This is because a muscle fiber is a long cylindrical muscle cell. You might also notice multiple dark, purplish spots inside the cells. These are the nuclei. Skeletal muscle cells are multinucleated, meaning that they have many nuclei in one cell. Muscle cells also have a lot of mitochondria. Do you remember what the role of mitochondria is in the cell? Yes, they produce the energy that cells need to carry out their basic functions. The main function of a muscle cell is to contract or shorten. This takes a lot of energy, which means the muscle cell needs a lot of

## Dark Meat or Light Meat?

When it is time to serve the turkey at Thanksgiving dinner, you may be asked if you prefer dark meat or light meat. The light meat is found in the turkey breast, while the meat found on the legs or thighs are a darker color.

The difference in color has to do with the different functions of these muscle groups. The turkey breasts are muscles that move the bird's wings. Turkeys do not fly much, so these muscles are used only when they are needed, like when the bird is trying to escape from a hunter or predator. They can contract very quickly, but they also tire quickly because they have low amounts of mitochondria and a smaller blood supply, resulting in a lighter color (less blood = lighter color). We sometimes refer to them as "fast-twitch" muscles because when the turkey moves them, they move fast and quick (picture a bird taking off by flapping its wings).

The thighs of the turkey, meanwhile, are used all the time because the turkey primarily stands and walks. These muscles must be able to work for long periods of time without becoming tired. As a result, they have high amounts of mitochondria and a rich supply of blood. This is what gives the thighs a darker color. We sometimes refer to them as "slow-twitch" muscles because their movements are slower and more methodical.

So next Thanksgiving, when someone asks you which sort of turkey you prefer, throw your mom or dad or aunt or uncle for a loop and reply, "I'll have some fast-twitch muscles, thank you!"

LIGHT MEAT

DARK MEAT

mitochondria (needs a lot of energy, just as you need a lot of energy to play outside). Unfortunately, you can't see mitochondria using a typical light microscope because they are too small. However, you *can* see that the muscle cells have a striped appearance. Because of this, we say that skeletal muscles are striated, meaning they have long and thin parallel streaks. The striations are caused by repeating patterns of rod-like muscle protein fibers in the cell called **myofibrils**.

There are different types of myofibrils found in muscle cells. Here we will look at two of them: **myosin** and **actin**. Myosin is a golf-club shaped protein filament. One end is long and thin, like the handle of the golf club. On the other end is a knob or "head" that protrudes out. In a muscle cell, many of these myosin filaments are wound together, resulting in a thick strand of myosin with many myosin heads poking outward. The heads are found on the two ends of the myosin, but not in the middle. Actin, meanwhile, is a thin protein filament. In the muscle cell, two strands of actin twist around each other. This twisted strand of actin is attached to a structure which we call the Z line because it zigzags.

Let's look more closely at one unit of actin and myosin filaments in the muscle cell. We call this unit the **sarcomere**. The sarcomere represents the space from one Z line to the next Z line. There are many actin filaments that are attached to the Z line, but they do not stretch the entire length of the sarcomere. In between the actin filaments you see the thick myosin. The myosin

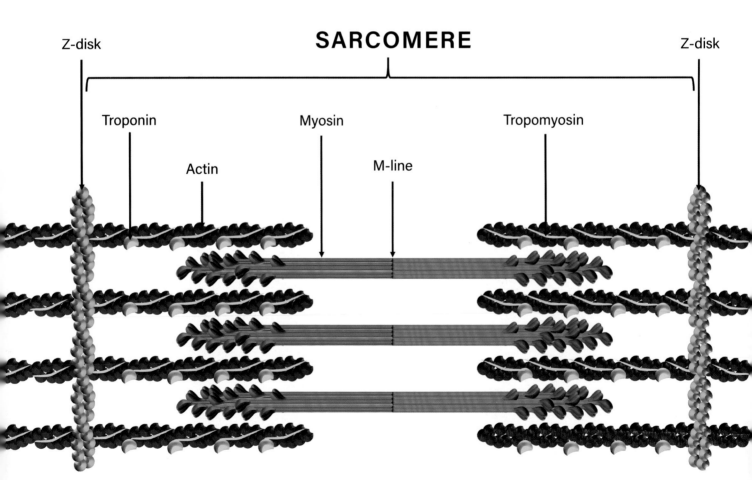

heads on either end of the myosin attach to the actin. This attachment is called a cross-bridge. Each muscle cell contains many sarcomeres lined up end to end. These repeating units result in the striations observed in skeletal muscle cells we spoke about a moment ago.

## MUSCLES CONTRACT

Muscles contract when they are stimulated or activated by a nerve cell. We will learn more about nerve cells in our next chapter. When a muscle contracts or shortens, this is because all the sarcomeres in the muscle fibers are shortening. How this happens is a truly an amazing feat!

For the sarcomere to shorten, the myosin heads pull on the actin filaments. After one pull, the myosin head let's go of the actin and reattaches at a different place on the actin so that it can pull again. You can think of this kind of like how a person rows a boat. The rower puts the oars in the water in front of him. Then he pulls on the oars, propelling the boat forward. The rowing motion leaves the oars behind, so he has to lift the oars up out of the water and move them forward so he can take another stroke. The repeated strokes move the boat forward. It is similar with the myosin heads. With all the myosin heads pulling repeatedly, the actin filaments are pulled closer to each other. Or another way to think about it is simply to say that the Z lines move closer to each other. This shortens the sarcomere. Of course, all of this pulling requires a lot of energy, just like rowing a boat takes a lot of energy. This is why the muscle cells need so many mitochondria (energy-producers!).

*Remember:*
*Muscles can either be voluntary or involuntary. Voluntary means we must think to move them, like flexing an arm. Involuntary means it moves and constricts without us thinking about it, like the muscle tissues that make up the heart.*

As the sarcomeres in a muscle shorten, the entire muscle shortens, or contracts. Once the muscle is done contracting, the myosin stops pulling on the actin and the sarcomere relaxes. Interestingly, the myosin heads can only pull. They cannot push. So when you contract or use your muscles, they can only shorten or pull.

## MAJOR MUSCLES

The "Muscles in Action" section introduced us to two muscles in your upper arm: the bicep and the tricep. Now let's introduce some of the other major skeletal muscles in the human body. Remembering their exact names is not as important as just getting a feel for where they are and what they do. As you read through this section, poke around and feel for them.

### Muscles in Action

To better understand how your muscles help you move, let's look at the example of the bicep muscle. Start by standing with your arms hanging down to your sides. The bicep in your upper arm is relaxed. Now, bring your hand up to touch your shoulder (the hand to the same side shoulder). The bicep contracted to pull your hand up. If you want to put your hand back down, you will have to use a different muscle. On the back of your upper arm is a muscle called the tricep. When the tricep contracts, it will pull your hand back down to its starting position. These two muscles, the bicep and tricep, work together to allow you to move your hand up and down. Muscles that work opposite each other are called **antagonistic pairs**. There are many examples of antagonistic pairs of muscles in the body that allow you to move in all different directions.

By the way, to help you remember this, an "antagonist" in a story is the character—the bad guy—who opposes the "protagonist," who is the main character you root for—the good guy. So these are called antagonistic pairs because they oppose each other, the way the bad guy opposes the good guy (although of course muscles don't have any agency or any moral intentions toward good or bad).

**Deltoid**: The deltoid muscle is found at your shoulder. Put your right hand on your left upper arm, just below your shoulder. Now lift your left arm straight up to the side. You may be able to feel the deltoid muscle tighten as it contracts to lift your arm.

**Pectoralis major**: These are the chest muscles. If you were to do a push-up, you would be using the pectoralis major muscles.

**Abdominal muscles**: This is the group of muscles found along your stomach. They contract when you do sit-ups.

**External obliques**: These are a set of large muscles found along the side of your upper body. They help you twist as well as bend side-to-side.

**Latissimus dorsi**: These are the big muscles found on your back. If you run your hands up either side of your spine, you may feel some of them.

**Gluteus maximus**: These are the muscles that make up the buttocks. They help you move your hip and thigh.

**Quadriceps**: This is a group of four muscles on the front of the thigh that help you lift your leg in a forward direction, which of course helps you walk and run.

**Hamstring muscles**: A group of muscles found on the back of the thigh. These work to pull your leg behind you (they are the antagonistic pair to the quadriceps).

**Gastrocnemius**: This is also known as the calf muscle. This muscle contracts when you stand on your toes.

**Tibialis anterior**: This muscle is found on the front of the leg. It hurts when someone gets a "shin splint." It is the antagonistic pair to the gastrocnemius.

# MUSCULAR SYSTEM
## *Voluntary Muscles*

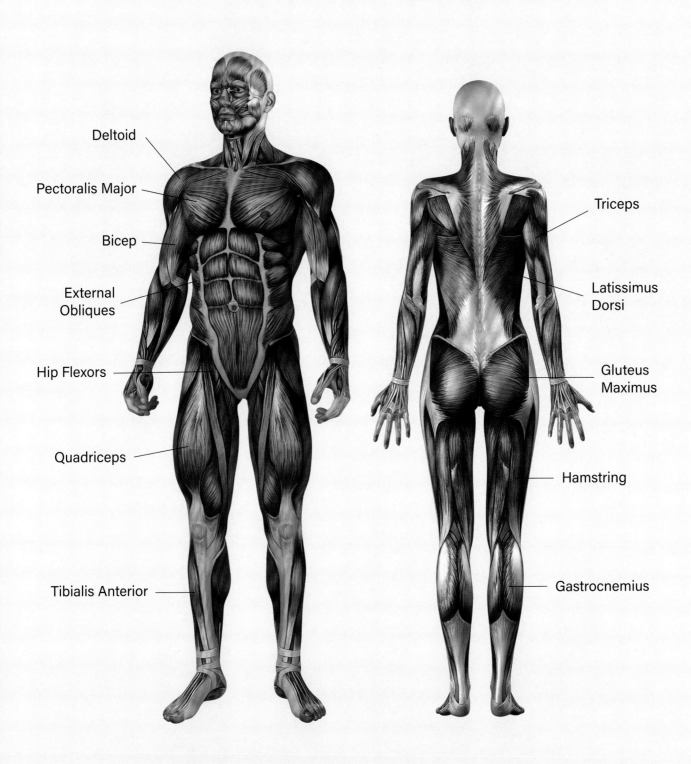

*Muscular System Fun Fact:*

*Muscles appear to bulge out and get bigger when you flex or use them. This is because the temporary tension causes the muscles to contract and "bunch up," making them denser. This is similar to squeezing part of a balloon, packing more air into a smaller, tighter area, causing it to swell up and bulge.*

## MAINTAINING MUSCLES

Our muscles are truly amazing organs, and as you have learned in this chapter, they are very important for you to be able to move. They also help to keep your heart beating, help you breathe, and move food through your digestive system. Therefore, it is important to keep your muscles in good condition. Doing so can help keep you from getting hurt. For example, people who have strong leg muscles are less likely to twist an ankle.

Healthy eating is very important for maintaining strong muscles. Because muscles are made mostly of protein, a diet that includes a lot of lean proteins such as chicken, beans, or eggs provides the proteins needed to build strong muscles.

Staying active is also important for keeping your muscles healthy. Your muscles get stronger when you use them. If you just sit on your couch all week, then your muscles start to get weaker because they are not being used. This is because the muscle fibers start to break down. We say that the muscle **atrophies**. So, once you finish your schoolwork for today, make sure you go outside and climb a tree, jump some rope, or ask your parents to take you to the playground. It's good for your muscles and bones!

### FOUNDATIONS REVIEW

✓ The human body has over six hundred muscles. These can be broken down into voluntary and involuntary muscles. Voluntary muscles are those which we can think about and move, while involuntary muscles are those which move without us thinking about them, like those in our heart or digestive system. Skeletal muscles help us move by contracting, or shortening, because when they do this, they move our bones.

✓ Most muscles are made up of muscle fibers held together in bundles by connective tissue. Muscle protein fibers in the muscle cells called myofibrils are what help the muscles contract, or move. Two types of myofibrils—myosin and actin—work together to create a unit known as a sarcomere. When a muscle contracts or shortens, this is because the myosin and actin are pulling on each other, shortening the sarcomeres, which shortens the muscle.

✓ Healthy eating and exercise help keep our muscles strong. It is important to keep muscles strong because if we don't, they start to atrophy, which means they break down and become weak. Because muscles are made mostly of protein, a diet that includes a lot of lean proteins such as chicken, beans, or eggs provides the proteins needed to build strong muscles.

# Strengthening Your Spiritual Muscles

*"Prayer is a pasturage, a field, wherein all the virtues find their nourishment, growth, and strength."*
—St. Catherine of Siena

A virtue is a characteristic or behavior that is seen as being morally good. Within our Catholic faith we can identify different types of virtues. The three theological (or supernatural) virtues are those that are given to us by God through supernatural grace. These are the virtues of faith, hope and charity (or love).

There are also four cardinal virtues: prudence, justice, fortitude, and temperance. Unlike the supernatural virtues, the cardinal virtues are acquired through practice (though of course all our gifts come from God originally). This means that if you do not practice using the virtues of prudence, justice, fortitude, and temperance, then they will become weak. This is similar to muscles. When muscles are not used regularly, they become weak and atrophy.

Prudence is the ability to determine what is good and what is bad. In a way, it's our ability to make good decisions. Justice, meanwhile, means to give to another what is due to them. For example, if you say that you will give your little sister a nickel if she lets you have a piece of her candy, then when she shares her candy, you owe her the nickel. The virtue of fortitude refers to one's ability to endure suffering or hardships and persevere when things are difficult (it is somewhat similar to courage or bravery but in spiritual battles). Finally, temperance is like moderation. We practice temperance when we do not eat all our Easter candy at one time, or when we choose to limit how much time we spend watching television.

Remaining close to God and living out the Church's teachings will help us learn to practice all these virtues, and with more practice, they will grow stronger in our soul, just like our physical muscles grow stronger when we use them.

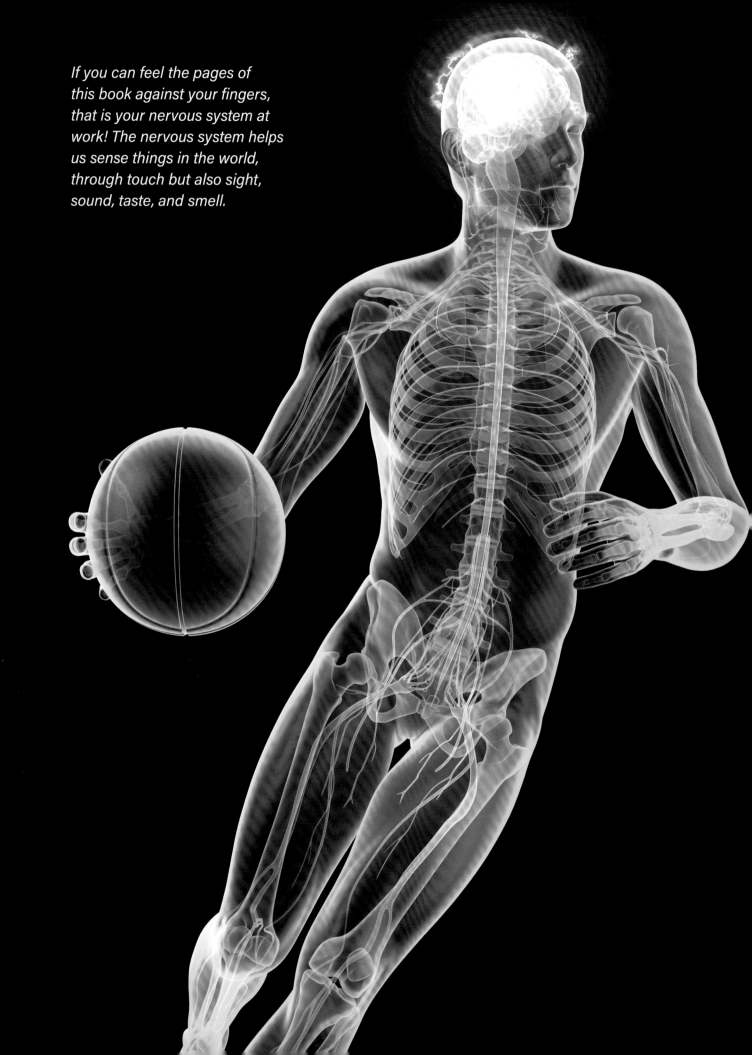

*If you can feel the pages of this book against your fingers, that is your nervous system at work! The nervous system helps us sense things in the world, through touch but also sight, sound, taste, and smell.*

# CHAPTER 7

## THE NERVOUS SYSTEM

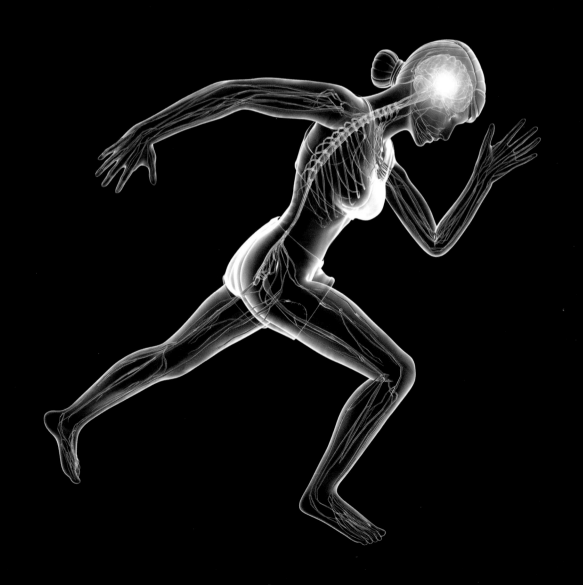

*Nervous System Fun Fact:*
*The human brain contains over eighty billion neurons!*

## SENSING THE WORLD

Have you ever had the chance to visit the beach? On a warm summer day, the sand feels warm, maybe even hot, underneath your bare feet. What do you usually do in this situation? Run to the water to cool them off of course! After this, you might smell the salty air and feel the warm breeze blowing against your face. You hear the waves crashing and the seagulls calling. You look out over the ocean and are amazed by how big and beautiful it is. All these sights, sounds, smells, and thoughts are made possible by your **nervous system**.

The nervous system allows us to *sense* things in our world, register them, and then respond in an appropriate way. For example, if you happen to be walking through your room and step on a LEGO, the nerves in your foot sense the sharp edges of the LEGO and send a message telling you to lift up your leg. But to understand how the nervous system can carry out these feats, we need to first have a better understanding of the cells that make up the nervous system.

## NETWORK OF NEURONS

**Neurons**, or nerve cells, are the main cells that makes up all the structures in the nervous system. A neuron has three parts to it. The **cell body** contains all the different cell organelles you learned about in chapter 2, including the nucleus. Extending off the cell body are numerous fingerlike extensions called **dendrites**. The dendrites pick up "information" from other cells or from the environment (like if something is hot or cold). A single long fiber that extends off the cell body called an **axon** carries this information away from the cell body and passes it on to other cells or to a muscle. Perhaps you can think of the axon like a long cable that carries electricity, like the power lines you see around your community. Some neurons have supporting cells that wrap around the axon to form the **myelin sheath**. This provides insulation

# NEURON

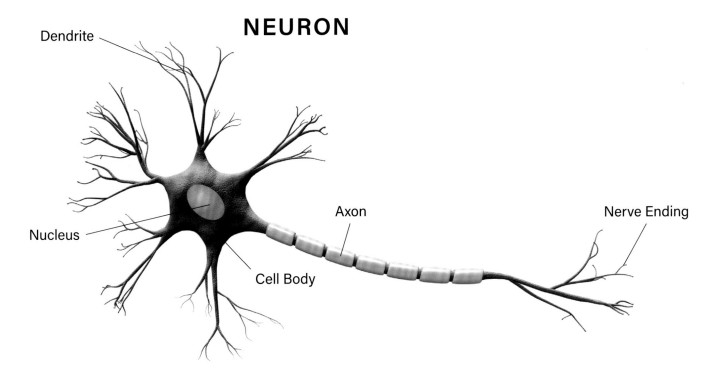

(protection) to the axon, allowing nerve signals to travel more quickly; receiving information from the environment quickly is important. (Don't you want to know as quickly as possible if something is too hot to touch?)

Neurons work together to carry information from one part of the body to another. Neurons that sense information from the environment are called **sensory neurons**. You find sensory neurons in your skin to sense pain, heat, or touch. There are also special sensory neurons in your eye that sense light, and in your ear to sense sound vibrations. The sensory neurons send information to an interneuron, which then passes the information to a motor neuron. **Motor neurons** carry information to a muscle or other organ that responds to the stimulus. Going back to the example of the LEGO, the sensory neuron is what feels the sharpness of the toy beneath your foot, and the motor neuron sends a message to the muscles in your leg to contract so that you lift your foot. In this way, neurons (nerve cells) work together like a colony of ants might all work together, each taking up different tasks.

A **nerve** is a group of neurons bundled together. These run throughout your entire body forming a network, allowing the different parts of the body to communicate with each other. It might help to think of this network like a circuit board in a computer that sends information all over the computer and helps the various parts work together. Nerves can also come together to form special nervous tissue and organs. The most prominent organ is the brain, which is estimated to contain over eighty billion neurons!

With this basic understanding behind us, let's discuss the different types of nervous systems in your body.

## The Largest Cell on Earth

The largest cell we know of is a neuron from a Giant Squid. This nerve stretches from the head of the squid all the way down to the end of its tentacles, making it the length of the squid, or about forty feet long! The axon of the neuron has a diameter of about one millimeter, making it very thin, but it can still be seen without using a microscope.

*Remember:*
*A nerve is a group of neurons bundled together. These run throughout your entire body forming a network, allowing the different parts of the body to communicate with each other.*

## CENTRAL NERVOUS SYSTEM

The nervous system is divided into two parts, the central nervous system and the peripheral nervous system. The **central nervous system** includes the brain and the spinal cord, while the **peripheral nervous system** includes all the other nerves in our body. Let's start with the central nervous system since the brain is like the computer or engine that runs everything else.

We divide the brain into three basic parts. The largest part of the brain is the **cerebrum**. From the outside, it looks kind of like a very large walnut. It has many grooves on its surface, with one large groove down the middle. The large groove divides the brain into two halves, or hemispheres. The main part of the cerebrum is divided into four lobes: frontal, parietal, occipital, and temporal.

## ANATOMY OF THE BRAIN

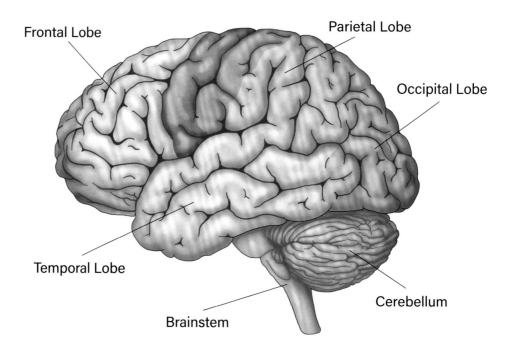

The frontal lobe of the brain is responsible for most of our conscious thoughts as well as skeletal movements. The parietal lobe mainly controls our ability to sense touch. Other parts of the cerebrum are associated with the senses of sight (occipital lobe), hearing (temporal lobe), taste, and smell. Other regions play roles in forming memories, learning, and communication.

The second part of the brain is the **cerebellum**. This is a smaller structure found underneath the occipital lobe. The main role of the cerebellum is balance and helping to coordinate our movements. The third part of the brain is the **brain stem**. This connects the brain to the spinal cord. You can think of the brain stem kind of like the stalk of a mushroom, where the cerebrum is the cap. The brain stem passes information between the brain and the spinal cord to send messages to and from the rest of the nervous system. It also contains special regions of neurons that regulate involuntary functions such as maintaining breathing and keeping your heart beating.

The other part of the central nervous system is the **spinal cord**. It is contained inside the vertebrae that make up the backbone, which you can feel if you run your hand down the center of your back. These bones protect the spinal cord from damage. In an adult, the spinal cord is about eighteen inches long and has a diameter of about one centimeter. Over forty pairs of nerves extend off the spinal cord and carry information to all the different parts of the body. This information is then sent back to the brain. In some cases, the spinal cord does not need input from the brain about how to respond. This is called a **reflex**, or an involuntary response to a stimulus. Reflexes, like the one we described when you step on the LEGO, help us to respond quickly.

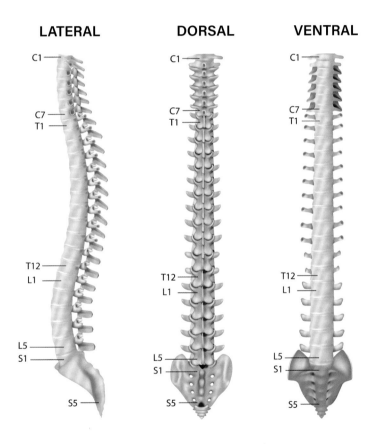

*The vertebrae column seen from three different angles. Medical experts divide it up into sections known as Cervical, Thoracic, Lumbar, and Sacral, with corresponding numbers running down (ex. C1, C2, etc.). Over forty pairs of nerves extend off the spinal cord (protected within the vertebral column) and carry information to all the different parts of the body.*

## PERIPHERAL NERVOUS SYSTEM

Now let's talk about the peripheral nervous system. Do you know what "peripheral" means? It's a word we use to describe something that sits on the edge, or outside, of something else. All the nerves that branch off the spinal cord and brain stem are part of the peripheral nervous system because they run to the outside edge of our bodies. There are hundreds of millions of nerves in our body. If you put them all together, they would reach around forty-five miles in length! Let's take a closer look at a few of them.

*The ulnar nerve is what is affected when we say you hit your "funny bone," but it is actually not a bone and it's not too funny!*

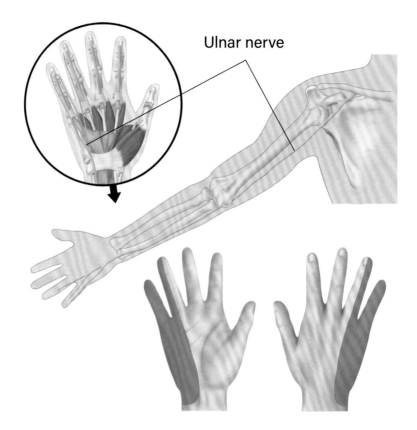

Ulnar nerve

*Ulnar nerve*: The ulnar nerve runs down your arm, around your elbow and into your hand. If you bump your elbow, you may accidentally hit the ulnar nerve, causing a tingling or numbing in your arm. We often say that you hit your "funny bone." It's an ironic name considering it is a nerve and not a bone, and since it hurts, hitting it is anything but funny!

*Sciatic nerve*: The sciatic nerve is the longest and thickest nerve in the human body, running from the lower back all the way down the leg to the foot. It is not uncommon for people to experience sciatic pain as they get older. One cause of this could be that the nerve is getting pinched between some of the vertebrae in the backbone.

*Phrenic nerve*: The phrenic nerve comes from the brain stem down to a muscle called the diaphragm. We will learn more about the diaphragm in chapter 9. It separates the chest region from the abdominal or stomach region. It is also very important in helping you to breathe. If the phrenic nerve gets irritated, it may cause the diaphragm to contract, resulting in hiccups.

## SPECIAL SENSES

So far we have learned a lot about the nerves that make up important nervous structures like the brain and spinal cord. But there are other organs that are also part of the nervous system. The special sense organs—such as the eye, ear, nose, and tongue—all contain special sensory neurons that help us sense stimuli in the world around us. Let's look at a few of these.

## Eye

The eye is made of many different structures that work together to allow us to see, including the **cornea**, where light first passes through. This is the clear surface that covers the outside of the eye. As light moves through the cornea, it bends a little and is directed through the **pupil**. The pupil is the black circle you see when you look at someone's eye (go look in the mirror at your own!). It is really an open window that allows light to enter the inside of the eye. The **iris**, or colored tissue part of the eye, is made of muscles that control the size of the pupil. If you are in a dark room, the pupil gets bigger to let more light in, but if you go outside on a bright sunny day, the pupil shrinks so that less light enters the eye. Next, the light will pass through the **lens**, a gelatin-filled sphere held in place by small muscles inside your eye. Those muscles can stretch the lens flat when you are looking at something far away, but when you look at something close, the lens gets fat and round. The lens changes shape to help focus the light on the back of the eye. The inner layer of the eye is called the **retina**. It contains the special sensory cells called **photoreceptors** that respond to light. Once the photoreceptor cells are activated, they send the message to other neurons and eventually to the **optic nerve**, which carries the message to the occipital lobe of the brain.

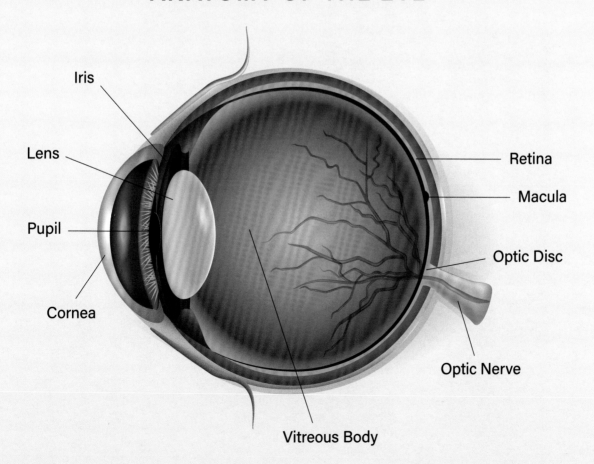

### Ear

Usually when you think of the ear, you think of those fleshy things attached to the sides of your head. But this is only one part of your ear, the **auricle**. The ear is divided into three parts, the outer, middle, and inner ear. The purpose of the auricle is to direct sound waves into the **ear canal**, the small tube that opens into the ear. Once inside, the sound waves hit a thin **tympanic membrane**, or eardrum. Together, the auricle, ear canal, and tympanic membrane make up the outer ear. Just like when a drummer strikes a drum, the membrane vibrates when it is struck by sound waves. Those vibrations are passed through three very small bones found inside the ear, the malleus (hammer), incus (anvil) and stapes (stirrup). The part of the ear that contains these bones is called the middle ear. The stapes has an oval shaped base that covers an opening in the cochlea called the oval window. The inner ear contains the **cochlea**, a snail-shell shaped structure filled with fluid. The vibrations from the stapes are passed to the fluid, causing small waves to form inside the cochlea. The waves bend small sensory cells inside the cochlea, causing them to send a signal through nerves in the inner ear. The nerves join to make the **auditory nerve**, which carries the signal to the temporal lobe of the brain.

## ANATOMY OF THE EAR

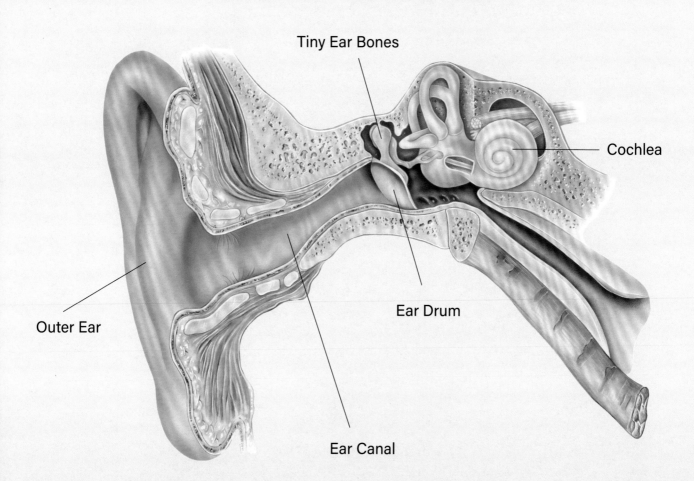

### Tongue and Nose

The sense of taste and smell are similar because they are both stimulated by chemicals. Chemicals in the air, odorants, enter the nose and attach to special sensory cells called **chemoreceptors**. The cells are activated by the chemicals and send a signal along the olfactory nerve to our brain. Meanwhile, taste occurs in the tastebuds on our tongue. The tastebuds also contain chemoreceptors. When sugar, salt, or other chemicals bind to the chemoreceptors, they are activated and send a signal through nerves to our brain. The sense of taste is largely affected by our sense of smell. If you ever had a cold with a stuffy nose, you probably noticed that your food does not taste as good because you can't smell. Now you know why!

These special sense organs work with the rest of the nervous system to bring in important information about the world around you. Hopefully now that you have read through this chapter, you can appreciate how we are able to enjoy the world around us through the nervous system that God created.

*Nervous System Fun Fact:*
*Our sense of taste is largely affected by our sense of smell.*

### FOUNDATIONS REVIEW

- ✓ Neurons, or nerve cells, are the main cells that make up all the structures in the nervous system. They work together to carry information from one part of the body to another. Neurons that sense information from the environment are called sensory neurons (sensing heat, cold, pain, light, sound, etc.). These stimuli are then sent to motor neurons, which carry information to a muscle or other organ that responds to the stimulus (removing your hand from a hot stove).

- ✓ The nervous system is divided into two parts: the central nervous system and the peripheral nervous system. The central nervous system includes the brain and the spinal cord, while the peripheral nervous system includes all the other nerves in our body.

- ✓ There are other organs that are also part of the nervous system. The special sense organs—such as the eye, ear, nose, and tongue—all contain special sensory neurons that help us sense stimuli in the world around us.

# Jesus Heals the Blind Man

Throughout the Gospels, we see many accounts of Jesus healing people who were blind or deaf. In the account according to Saint John, He heals a man who was born blind. Jesus's disciples asked Him if the man was born blind because of his sin or because his parents sinned. In those days, people thought that suffering and misfortune was a sign of having angered God and He was punishing you. We know that is not the case now thanks to the theological teachings of the Church on human suffering and, in large part, because of this very story we are discussing. Jesus clarifies that the man was born blind not because of some sin he or his parents committed but so that God could be glorified. And the disciples witness this glorification of God when Jesus gathers up mud and puts it on the man's eyes, then tells him to go wash in the pool of Siloam. After washing the mud off, the man is able to see! Jesus gave him complete healing from the blindness that he had since birth. In this

specific instance, Jesus used the man's blindness to manifest His divinity. God can also use our sufferings to manifest His glory in so many ways.

Just as Jesus healed this man of his physical disease, so He also heals us of our spiritual diseases that are the result of sin. Because of the original sin of Adam and Eve, we also are born blind. But our blindness is a *spiritual* blindness. Through the sacrament of Baptism, we are cleansed of this stain of sin and restored to the state of grace. This allows us to see the light of Christ with our spiritual eyes. The Bible makes many references to living in the light. In John 8:12, Jesus says, "I am the light of the world; he who follows me will not walk in darkness but will have the light of life." Each day, we get to make the choice to follow God and walk in the light.

But we could also choose not to follow God. This will lead us into darkness and spiritual blindness, making it difficult to live a life of grace and leading us into a life of sin. Sin clouds our vision and darkens our ability to see the light of Christ. We know from our catechism that even small venial sins weaken the life of grace within us and lead us away from God. Through the sacrament of confession, we can once again cleanse our soul and restore our spiritual sight. This allows us to continue to live in the light of Christ.

Saint Lucy of Sicily is the patron saint of the blind and those with eye problems. Her mother was healed of blindness through the intercession of Saint Agatha. Out of gratitude, Saint Lucy decided to remain a virgin like Saint Agatha, but she had been betrothed to marry a pagan. When she refused, she suffered many torments. It is believed that one of these torments was having her eyes removed. Tradition suggests that when she was buried, her eyes had been restored and replaced. For this reason, she is often pictured carrying a platter with two eyes.

*O Saint Lucy, preserve the light of my eyes so that I may see the beauties of creation, the glow of the sun, the color of the flowers, and the smile of children.*

*Preserve also the eyes of my soul, the faith, through which I can know my God, understand His teachings, recognize His love for me, and never miss the road that leads me to where you, Saint Lucy, can be found in the company of the angels and saints. Amen.*

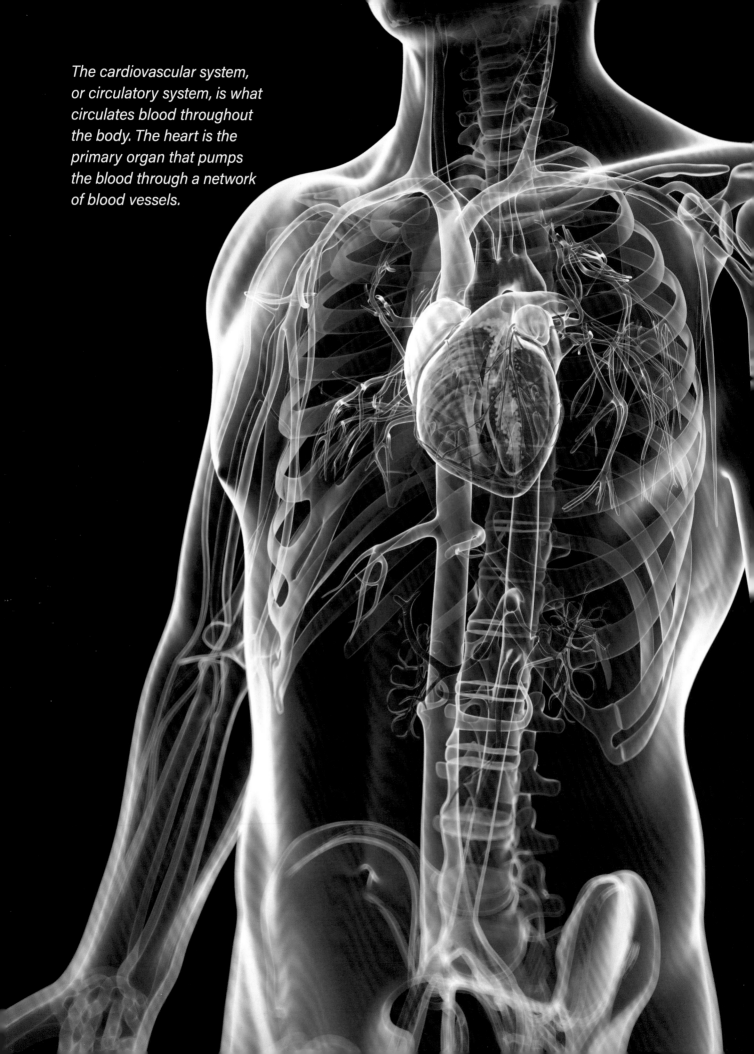

The cardiovascular system, or circulatory system, is what circulates blood throughout the body. The heart is the primary organ that pumps the blood through a network of blood vessels.

# CHAPTER 8

## THE CARDIOVASCULAR SYSTEM

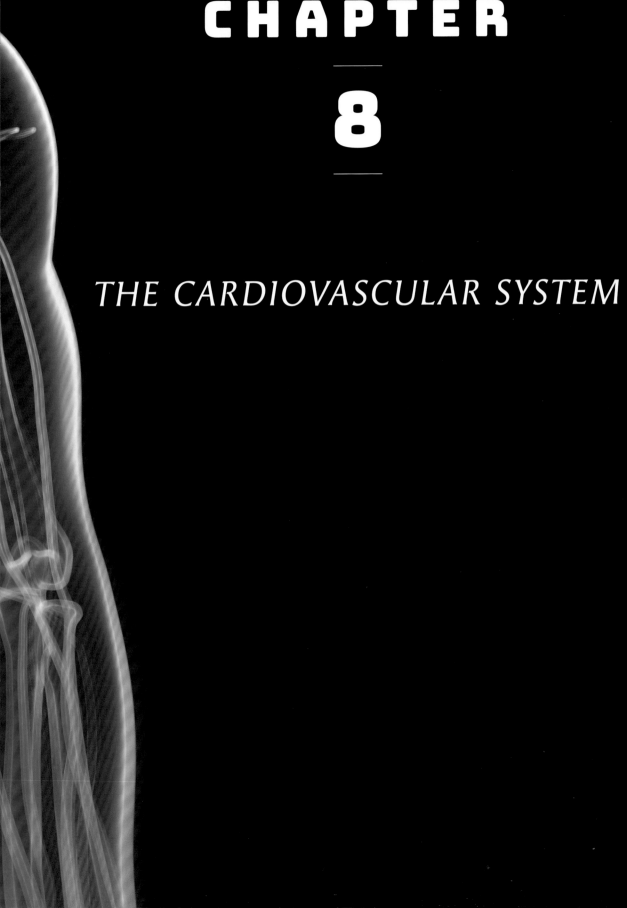

## THE SACRED HEART

The heart is often associated with ideas of emotions, feelings, and love. When someone is in love, we use the heart as a symbolic way to show this love. If people hurt our feelings, we sometimes say that they broke our heart. Even in our Catholic faith we see this imagery portrayed through the devotion to the Sacred Heart of Jesus. In the seventeenth century, a nun by the name of Saint Margaret Mary Alacoque received several visions of Jesus asking to be honored by the symbol of His Sacred Heart.

The feast of the Sacred Heart was added to the liturgical calendar in 1856 to be celebrated on the Friday following the second Sunday after Pentecost. Jesus's heart burns with unbounding love for each of us, and we see this in the sacrifice He made for us on the cross at Calvary. But why is the heart used to show such strong emotions and feelings? Perhaps it is related to the important role of the heart in the physical body.

The heart is the central organ in the **cardiovascular system**, which is the focus of our present chapter. Sometimes you will hear the cardiovascular system referred to as the circulatory system. This is because its primary function is to circulate blood throughout the body. The heart obviously plays a huge part in this. Its main role is to pump blood through a network of **blood vessels**—tubes through which the blood circulates throughout the entire body—carrying oxygen and nutrients to all the different cells. The oxygen and nutrients are used by cells to make energy, which your body needs to do many things (walk, think, play, work, etc.). In the process, the cells produce waste products and a waste gas called carbon dioxide. These waste products are carried, once again by the blood, to be removed from the body. Let's take a closer look at each of these components.

*Cardiovascular System Fun Fact:*
*If you removed all the blood vessels (arteries, capillaries, veins) from an average-sized child and stretched them in a line, they would be over 60,000 miles long! For an adult, they would be even longer, nearly 100,000 miles!*

## BLOOD: THE SOURCE OF LIFE

Blood is made mostly of water and cells. An adult has about five liters of blood in his body—that's two and a half bottles of soda! The watery part of the blood is called **plasma**. Dissolved in the plasma are other chemicals such as salts, sugars, and proteins. Three different types of cells are suspended in the plasma, each having their own function:

**White blood cells** help our bodies to fight off infections. (Think of them like little knights that fight and ward off invading barbarians bringing sickness!)

**Platelets** are tiny fragments of cells that clump together to stop bleeding. These are important when you get a cut.

**Red blood cells** carry oxygen in the blood.

You probably know that blood is red. A protein called **hemoglobin** found inside the red blood cells is what gives it this color. Each hemoglobin protein carries four molecules of heme that hold an iron atom. It is the heme that

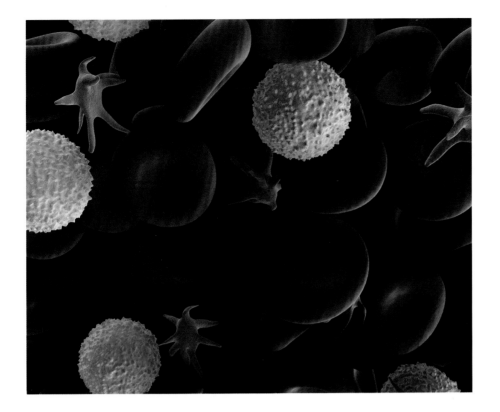

*When our blood is examined at a microscopic level, we can see white and red blood cells, as well as platelets (the ones that look like a starfish!), tumbling along in the current of our blood vessels.*

binds to an oxygen molecule and gives red blood cells their red color. Without iron, our body cannot produce the hemoglobin it needs to carry oxygen. Each red blood cell contains about 250 million hemoglobin molecules; this means that each cell carries about one billion oxygen molecules! The oxygen molecules enter the blood when it moves through the lungs. We will learn more about the lungs in our next chapter on the respiratory system.

## Blood Types

You may have heard that people have different "blood types." This doesn't mean the blood is different. It still has the same cells and chemical makeup. The difference comes from the types of proteins found on the outside of the red blood cell.

There are two types of proteins: A proteins and B proteins. Someone who has only A proteins on the outside of their red blood cells is said to have Type A blood, while someone with B proteins on the outside of their red blood cells has Type B blood. If a person has both A and B proteins, then they are said to have Type AB blood. Interestingly, other people may have neither protein. When this happens, we say that the person has Type O blood.

In a major surgery, if a patient needs to receive a blood donation (he is losing so much blood that he must have new blood pumped into him), he must receive blood with the same proteins. In other words, if you are Type A, you need someone with Type A blood to make the donation. If the wrong blood type is pumped into someone, it can cause a severe reaction and threaten his life. But there is an exception to this rule: *Anyone* can receive Type O blood because it is considered the universal donor, meaning everyone's body can receive it.

## BLOOD VESSELS: HIGHWAYS TO THE HEART

The blood in our bodies is only found within blood vessels. There are three main types of blood vessels: arteries, capillaries, and veins. Blood leaves the heart through a very large blood vessel called the **aorta**. This is the largest artery in the body. **Arteries** are blood vessels that carry blood *away* from the heart. They have thick walls because they must be able to withstand the pressure of the blood pushing against them as the blood leaves the heart. You can think of the arteries like major highways. They carry a lot of blood, just like major highways "carry" a lot of cars. The arteries branch into many smaller arteries called **arterioles**, just like a highway has small roads that branch off of it. The arterioles have thinner walls and get smaller and smaller until eventually they pass the blood into tiny blood vessels called capillaries.

**Capillaries** are the smallest blood vessels. They are so small that only one red blood cell can squeeze through at a time. The capillaries are found in all our muscles, skin, and tissues. It is in the capillaries that oxygen leaves the red blood

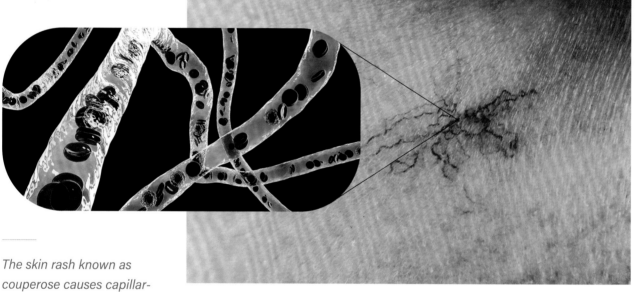

*The skin rash known as couperose causes capillaries to become dilated so that we can actually see them spread out in spider web-like red tones.*

cells, passes through the wall of the capillary, and moves into the surrounding tissues. At the same time, the carbon dioxide that was produced by the cells leaves the tissues and moves into the blood in the capillaries. After passing through the tissues, millions of capillaries join back together to form little veins. This is just like small roads merging together into bigger roads.

**Veins** are blood vessels that *return* blood to the heart. Small veins, called **venules**, merge together into bigger veins. The largest vein in the body is the **vena cava**, which dumps blood back into the heart.

## THE HEART OF THE MATTER

As we have learned, the heart is an organ, but it is made up of muscular tissue, and so it is sort of like a muscle that pumps blood through the blood vessels.

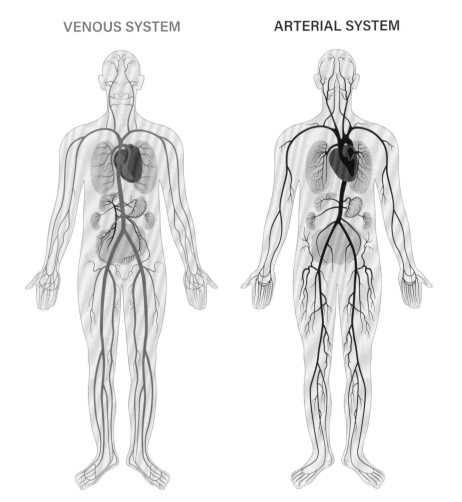

VENOUS SYSTEM   ARTERIAL SYSTEM

*Think of veins and arteries like a highway where traffic blows past one another in different directions. The veins (venous system) return blood to the heart, while the arteries (arterial system) take blood from the heart to the rest of the body. We can see the veins on the surface of our skin, but arteries are buried too deep to be seen.*

Since it has this muscular tissue, the heart contracts just like your other muscles, squeezing the blood out of its chambers (sections of the heart). There are four chambers in the heart. Two of the chambers *receive* the blood from the veins as they return the blood back to the heart. These are called **atria** (*singular* atrium). There is an atrium on the left side of the heart that receives blood returning from the lungs and an atrium on the right side of the heart that receives blood from the vena cava returning from the rest of the body.

The other two chambers are the **ventricles**. These are the chambers that *force the blood out* of the heart when they contract. The left ventricle sends blood out of the aorta to the entire body, while the right ventricle sends blood out of the pulmonary artery to the lungs.

Separating the atria and the ventricle are flap-like structures called **valves**. These are like one-way doors that control the movement of blood through the heart. Due to the valves, the blood can only move in *one direction*: from the atria to the ventricles. There is another set of valves that keeps the blood from moving back into the ventricles once it exits out into the arteries.

When we talk about the right and the left side of the heart, or any organ in the body, we refer to the right and left on the person. As you look at the image of the heart, pretend that you are in the picture and that is your heart. Notice

# ANATOMY OF THE HEART

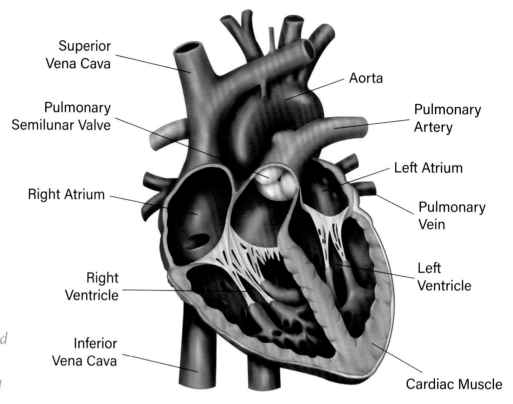

*Cardiovascular System Fun Fact: Sometimes people need a heart valve replaced because it has stopped functioning properly – doctors can replace it with a mechanical valve, or even a valve from a cow or pig!*

that the right side of the heart is colored blue, while the left side of the heart is colored red. We use the colors blue and red to show blood that has low or high levels of oxygen. Blood with high levels of oxygen is shown to be red, while blood with low levels of oxygen is shown to be blue. The blood on the right side of the heart has low levels of oxygen because it has just returned from the body where it gave oxygen to the muscles and tissues. It is going to be sent out to the lungs to pick up more oxygen. The blood on the left side of the heart is returning from the lungs. Thus, it has high levels of oxygen and is going to be sent out to the muscles and tissues to deliver the oxygen.

As the heart contracts, blood is forced out of both ventricles at the same time. Some of it goes to the lungs and some goes out of the aorta to be sent through the arteries to the muscles and tissues. The blood pushes against the walls of the arteries with each heart contraction. This results in a pulse that you can feel. Try finding your pulse by placing your three middle fingers on the inside of your wrist. Count how many times the heart beats each minute. This is your **heart rate**. While at rest, a child usually has a heart rate between eighty and ninety beats per minute. But when you play or exercise, your heart rate will increase (become faster). This is because the heart needs to move more blood through the blood vessels to deliver more oxygen to the muscles.

## PUTTING IT ALL TOGETHER: THE PATH OF BLOOD

Now that we have looked at all the different components of the cardiovascular system, let's put it all together to see how the cardiovascular system is able to move blood, and oxygen, to the entire body. Follow along as we trace the path of blood.

Pretend you are a red blood cell just returning to the heart from the body. The red blood cell enters the right atrium through the vena cava. It is then passed through the valve down into the right ventricle. When the heart contracts, the red blood cell is propelled out of the heart through the pulmonary artery. As it exits, it passes through the pulmonary valve. This carries it into a network of blood vessels in the lungs. While in the lungs, the red blood cell picks up a load of oxygen and gets rid of the excess carbon dioxide it was carrying. Now the red blood cell returns to the heart through the pulmonary vein. It enters into the left atrium and passes through the valve down into the left ventricle. This time when the heart contracts, the red blood cell is propelled out of the heart through the aortic valve and into the aorta. It moves through arteries, arterioles, and finally into capillaries. Here it must slow down so that it can unload the oxygen into the surrounding cells. While it is there, the red blood cell also picks up carbon dioxide. Now it moves into venules and finally larger veins until at last it makes its way back to the vena cava and into the right atrium of the heart. This entire journey is made in about forty-five seconds!

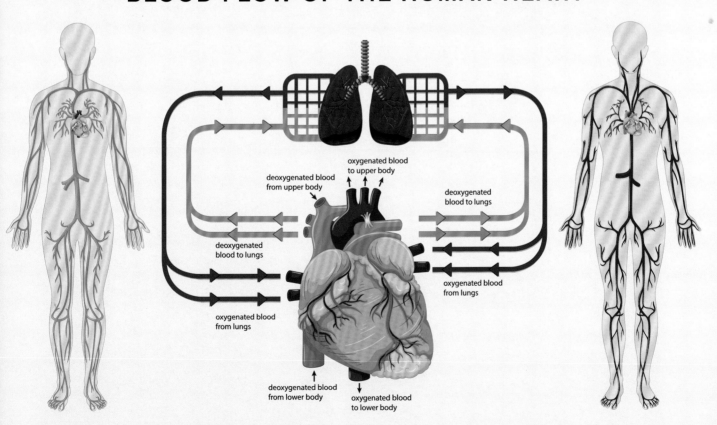

## BLOOD FLOW OF THE HUMAN HEART

## A HEALTHY HEART

Keeping our hearts strong and healthy is very important for our overall well-being. One way to have a healthy heart is to eat a well-balanced diet with fruits and vegetables, whole grains, and low-fat proteins. Limiting the amount of processed food and foods with large amounts of sugars, salts, and fats is also important for keeping your heart healthy.

Another important way to keep your heart strong is to use it. Because the heart is made of muscle, it must be used to get stronger. Specialists recommend doing moderate exercise for about twenty minutes a day. The exercise should increase your heart rate to 65–75 percent of your maximum heart rate. This is called the "target heart rate." Maximum heart rate is 220 minus your age. So if you are ten, your maximum heart rate would be 210 and your target heart rate would be 135–158 beats per minute. Now that you have finished reading this chapter, grab a heart-healthy snack and spend some time running and playing outside. Afterall, it's good for your heart!

## FOUNDATIONS REVIEW

✓ The heart is the central organ in the cardiovascular system. Its main role is to pump blood through a network of blood vessels—tubes through which the blood circulates throughout the entire body—carrying oxygen and nutrients to all the different cells. The oxygen and nutrients are used by cells to make energy, which your body needs to do many things (walk, think, play, work, etc.).

✓ Three different types of cells are suspended in the plasma of our blood: white blood cells, platelets, and red blood cells. White blood cells help our bodies to fight off infections. Platelets are tiny fragments of cells that clump together to stop bleeding. Red blood cells carry oxygen in the blood as well as a protein called hemoglobin that gives blood its red color.

✓ There are three main types of blood vessels—arteries, capillaries, and veins—through which blood moves throughout the heart, lungs, and the rest of the body. Arteries are blood vessels that carry blood *away* from the heart. Capillaries are the smallest blood vessel. They are where oxygen leaves the red blood cells, passing through the wall of the capillary, and moving into the surrounding tissues. After passing through the tissues, millions of capillaries join back together to form little veins. Veins are blood vessels that return blood to the heart.

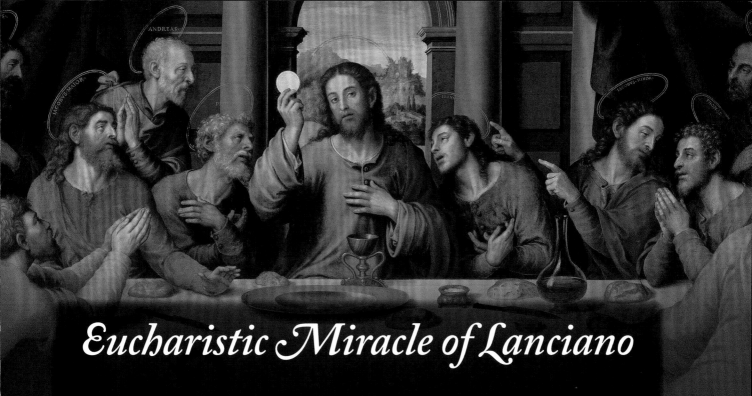

# Eucharistic Miracle of Lanciano

At the Last Supper, Christ filled the cup with wine and said, "Drink of it, all of you; for this is my blood of the covenant, which is poured out for many for the forgiveness of sins" (Mt 26: 27–28). As Catholics we believe that with these words, Christ transformed the wine into His blood.

The beauty of this mystery was revealed to a priest in the year 750 in the Italian town of Lanciano. This particular priest doubted the mystery that the bread and wine truly become the Body and Blood of our Lord. While he was celebrating Mass one day, at the words of consecration, the host changed physically into flesh and blood. The archbishop was called for to attest to the miracle and samples were placed in a reliquary, but were not sealed for preservation. Miraculously, even today, the flesh and blood that appeared in 750 have not rotted or deteriorated but remain in their original form.

Over the past several centuries many studies have been done on the flesh and blood of Lanciano, showing that they have the same properties of living human flesh and blood. Interestingly, it has a blood type of AB, the same type of blood identified in the Shroud of Turin (the burial cloth that is believed to have wrapped our Lord after His death). The flesh was analyzed under a microscope and has the same appearance of heart muscle tissue. What an amazing testament to the gift God has given us in the Eucharist. May we always receive it with reverence, humility, and thanksgiving!

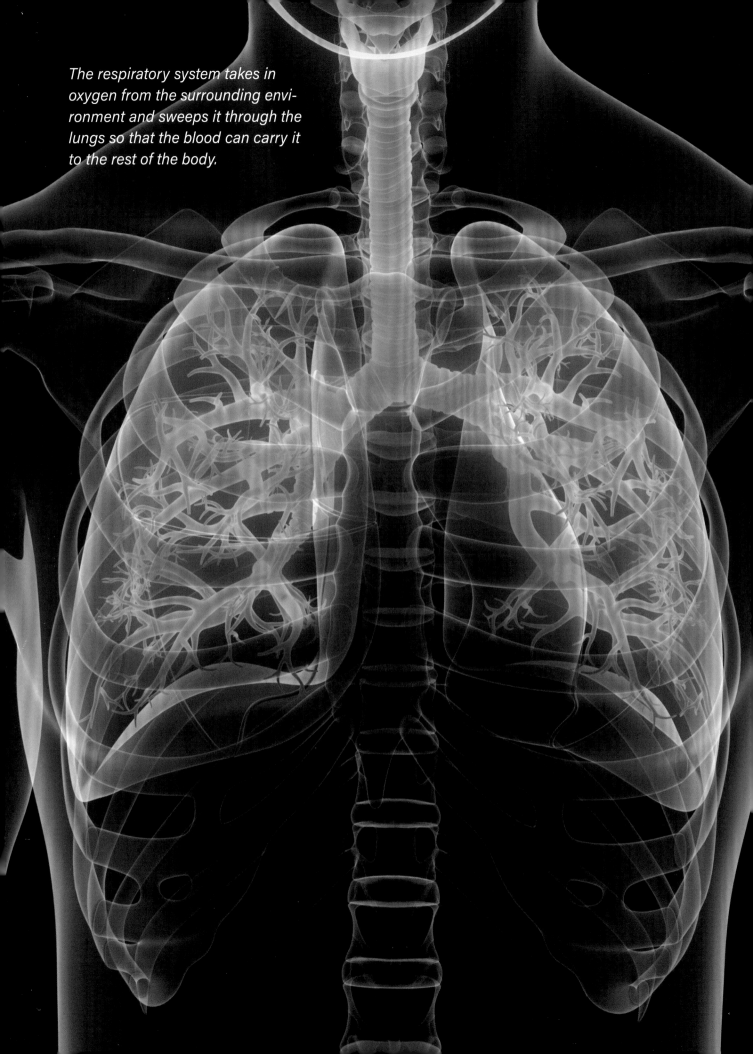

*The respiratory system takes in oxygen from the surrounding environment and sweeps it through the lungs so that the blood can carry it to the rest of the body.*

# CHAPTER 9

*THE RESPIRATORY SYSTEM*

## THE BREATH OF LIFE

"Then the Lord God formed man of dust from the ground, and breathed into his nostrils the breath of life; and man became a living being" (Gn 2:7).

We all know that we have to breathe air in order to stay alive. But why do we need to breathe and how does it work? In this chapter, we will answer these questions and more as we learn about the respiratory system.

The **respiratory system** includes your nose, mouth, throat, lungs, and other structures associated with them. Its main job is to take oxygen gas into your lungs so that it is available for the blood to pick up and carry to the rest of your body. (Remember what we learned in the last chapter on the cardiovascular system?) So take a deep breath and get ready to dive in!

## RESPIRATORY ANATOMY: THE PATH OF AIR

Before we begin, we should probably talk a little about the air that we breathe. The air around us is a mixture of gases. If you read the *Foundations of Science* unit on Earth, you may remember that most (~78 percent) of the air we breathe consists of nitrogen gas. The second most abundant gas is oxygen (~20 percent). When we breathe in air, we are breathing in the mixture of all the gases, but our bodies pull out the oxygen gas that we need to survive.

Air enters our bodies through our nose. Yes, you can also breathe in through your mouth (like when you have a stuffy nose because of a cold), but most of the time we breathe in through our noses. And this is better because when the air comes into your nose, it gets warmed up and moisture is added to it so that it will not damage the tissues that line your respiratory system. How does this happen? The skin inside your nose, or **nasal cavity**, contains thousands of little capillaries. The blood in the capillaries is warm and helps to warm up the air. The air becomes moist by the mucus (or snot) that lines the inside of your nasal cavity. When you breathe in through your mouth, the air does not get warmed or humidified. This is why when you have a cold and can't breathe through your nose, you end up having a sore throat. The cold, dry air dries out the back of your throat, causing it to be sore. The mucus is also very important because it traps particulates, bacteria, or other things that could cause you to get sick. This is one of the main reasons why we have mucus lining the entire respiratory system.

Next, the air moves from the nasal cavity down through the back of the throat, called the **pharynx**. The pharynx is shared by both the respiratory and digestive system. At the bottom of the pharynx, there are two openings. One opening is to the esophagus, which leads to the stomach. We will talk about it when we discuss the digestive system in our next chapter. The other opening moves air into the **larynx**, or voice box. The voice box is a tubular structure connecting the pharynx to the **trachea**, or windpipe. It contains several pieces of cartilage that hold it open. The largest is the thyroid cartilage. In adult males, it can easily be seen as the "Adam's apple" on the front of the neck. Gently

*Respiratory System Fun Fact:*

*When we cough or sneeze, we dispense from our bodies foreign and possibly harmful substances we have accidentally inhaled. Yawning also serves a purpose, bringing extra oxygen into the lungs when our brain senses a shortage.*

place your hand over your thyroid cartilage and then swallow. You may be able to feel it rise up. When we swallow, an internal spoon-shaped piece of cartilage called the **epiglottis** covers the opening to the trachea. This prevents food or drink from entering into the respiratory system.

Inside the voice box are folds of throat tissue that form the **vocal cords**. As air passes over the vocal cords, they vibrate to produce sound. The pitch—that is, the highness or lowness—of the sound depends on how tight the vocal cords are stretched. Vocal cords that are more tightly stretched produce higher sounds while vocal cords that are more relaxed produce lower sounds. People who use their voices a lot, like a singer or someone cheering loudly at a sporting event, may lose their voice or develop laryngitis. Laryngitis is inflammation or swelling of the vocal cords. The best remedy is to rest the voice so the vocal cords can heal.

So far we have covered the structures that make up the upper respiratory tract: nasal cavity, pharynx, and larynx. Now we will move into the structures of the lower respiratory tract, starting with the trachea.

*The respiratory system has over a dozen major parts that operate in a network together to take in, regulate, and circulate air throughout the body.*

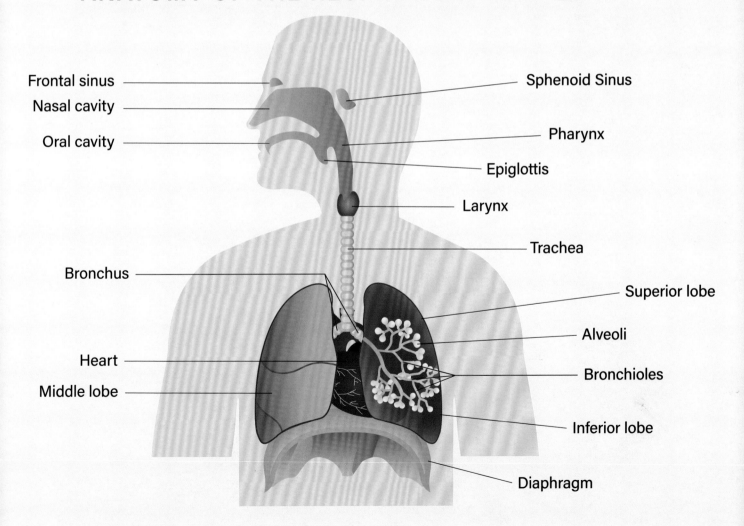

## ANATOMY OF THE RESPIRATORY SYSTEM

- Frontal sinus
- Nasal cavity
- Oral cavity
- Sphenoid Sinus
- Pharynx
- Epiglottis
- Larynx
- Trachea
- Bronchus
- Superior lobe
- Alveoli
- Heart
- Bronchioles
- Middle lobe
- Inferior lobe
- Diaphragm

The trachea, or windpipe, is a tube that connects to the base of the larynx. In an adult, it is about four to five inches long and has a diameter of between 0.5 to 0.75 inches. The trachea is held open by small, C-shaped rings of cartilage so that air can easily move through it. The trachea extends down into your chest and then divides into two smaller tubes called **bronchi**. One bronchus goes to the left lung and the other bronchus goes to the right lung. Once inside the lungs, the bronchi branch into smaller and smaller tubes called bronchioles. This is similar to what you see in a tree. Imagine that the tree trunk is the trachea. The trunk separates to form two main branches, which form more smaller branches, and so on. At the ends of the smallest bronchioles are tiny air sacs called **alveoli**.

Each alveolus is surrounded by a network of capillaries. As the blood moves through the capillaries, oxygen moves out of the alveoli and into the blood. At the same time, the carbon dioxide in the blood moves out of the blood and into the alveoli. While the alveoli themselves are very tiny (you can only see them under the microscope), if you were to take all the alveoli together and lay them out flat, they would cover most of a tennis court. This means that there is a lot of area for the oxygen and carbon dioxide gases to cross into and out of the blood. The oxygen is carried by the hemoglobin in the blood to the

*The anatomy of the respiratory system resembles a tree with a trunk splitting off into many branches, with the trachea dividing into the two bronchi that reach into the lungs, and from there splitting off into smaller tubes called bronchioles. The alveoli (air sacs) on the ends of the bronchioles can even be thought of like leaves on the end of branches.*

## DIAGRAM OF THE LUNGS

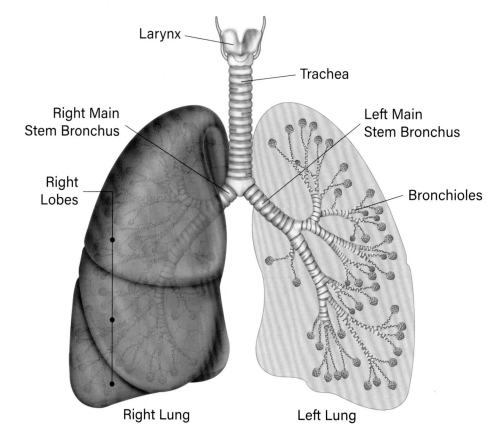

different tissues in the body. The carbon dioxide exits out of the respiratory system the same way the oxygen came in, only in reverse.

The bronchi, bronchioles, and alveoli are all inside the lungs. You have two lungs, and together they take up most of the space inside your chest cavity. The right lung is a little bigger than the left. This is because the heart sits between the two lungs but leans to the left side, making the left lung a bit smaller. Most of the volume taken up by the lungs is filled with air. This makes the lungs very spongy. In fact, lungs are the only organ in the human body that will float in water. The lungs sit on top of the **diaphragm**, a thin muscle that separates the chest cavity from your stomach cavity. The diaphragm plays a major role in breathing. But to understand how, let's shift our focus a bit.

## Asthma

Do you know someone who has asthma? **Asthma** is a medical condition of the lungs. The air passageways, such as the bronchi and bronchioles, constrict or get smaller. This makes it harder for air to move through

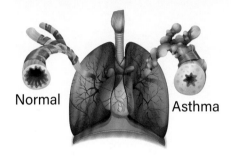

Normal    Asthma

them, causing the person to have shortness of breath, coughing, or wheezing. Scientists are not exactly sure what causes asthma. Some theories say that it could be caused by an allergic reaction to dust or pollens. It may also happen if a person has a strong emotional response to a situation. People who experience asthma regularly can get medicine from their doctor to help them breathe more easily. You have probably seen people use an inhaler. This has medicine in it that helps to open up the air passageways and expand the airways, making breathing easier.

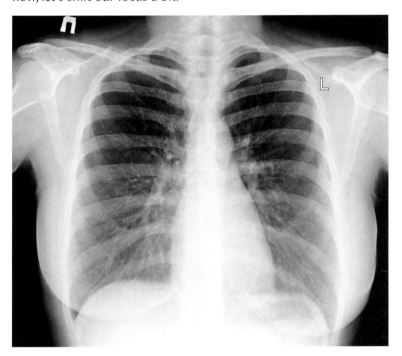

## VENTILATION: BREATHE IN, BREATHE OUT

Now that we know the path that air takes in and out of the lungs, let's talk about *how* it moves in and out of the lungs. **Ventilation** is the movement of air in and out of your lungs, just as the ventilation system of your house has to do with the air conditioning and heat moving through the systems and vents in your attic and floors and ceilings. It consists of two steps: inhalation, or bringing air into the lungs, and exhalation, moving air out of the lungs. (You are probably familiar with the terms "inhale" and "exhale.")

The primary muscle controlling ventilation is the diaphragm. When the diaphragm is relaxed, it curves upward. During inhalation, the diaphragm contracts and flattens out, increasing the volume of the chest cavity. The increase in chest cavity volume creates a slight vacuum, bringing a rush of air into the lungs. Exhalation occurs when the diaphragm relaxes, causing the volume of the chest cavity to get smaller and the air to move back out.

Try taking a deep breath. Did you notice that your chest went up when you inhaled? A second set of muscles that controls ventilation are the muscles between your ribs. These are called the **intercostal muscles**. When the intercostal muscles contract, they pull the rib cage out and up, making the chest cavity bigger. This is what happened when you took that deep breath. To exhale, the intercostal muscles relax and the rib cage falls back down, bringing the chest cavity back to its normal size again. The diaphragm and intercostal muscles work together to allow us to breathe in and out all day. Most people breathe between fifteen and twenty-five times in a minute, although children and women breathe a little bit faster than men. This is called the **respiration rate** and is controlled by the respiration center in the brain stem. A nerve from the brain stem sends a message to the diaphragm and intercostal muscles telling them to contract. When they contract, you take a breath.

We can control when we breathe. You just did this when you took a deep breath. You could also choose to hold your breath and not breathe. But our ability to control how often we breathe is limited. This is because the body is able to sense how much oxygen and carbon dioxide is in the blood. If there is too much carbon dioxide in your blood, special chemical sensors send a message to the brain stem. The brain stem then sends a message to the diaphragm telling it to contract so that you will breathe. This is what happens if you try to hold your breath too long. Your brain tells your body that you need to get rid of the carbon dioxide and forces you to take a breath.

*Our breathing is regulated by a two-step process of inhalation (bringing air in) and exhalation (moving air out). Try taking a deep breath (inhalation) without your chest cavity swelling up . . . it's not possible!*

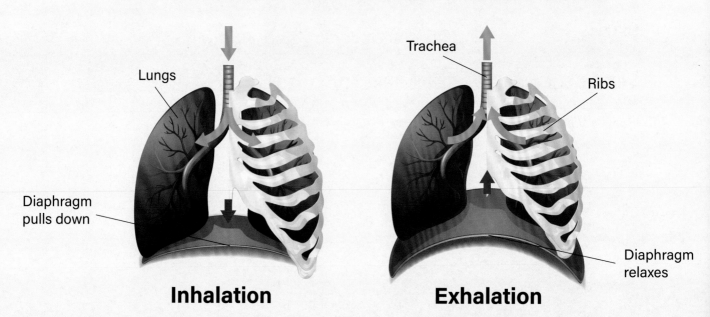

## TAKING CARE OF YOUR RESPIRATORY SYSTEM

Just like with your muscles and heart, exercise is also important for your lungs and respiratory system. It helps to strengthen your breathing muscles and makes your lungs more efficient, or better, at exchanging oxygen and carbon dioxide. Another thing that will help keep your lungs healthy is to avoid chemicals or other things you might breathe in that could damage your lungs. Smoking cigarettes, for example, is very bad for your lungs. The smoke irritates the lungs and causes the lungs to make too much mucus. The extra mucus clogs the airways in the lungs and causes a person to cough. If a person smokes a lot, it can cause long-term damage to the lungs by breaking down the alveoli. People who are around those who smoke a lot can also experience damage to their lungs from breathing in the "second-hand" smoke. Other chemicals, such as the fumes from paints or car exhaust, can also cause damage to your lungs. For this reason, it is also important to avoid breathing in chemical fumes.

You probably never guessed how complicated the simple act of breathing could be. Next time you're outside on a beautiful day, take a deep breath of fresh air and thank God for how wonderfully He has made you!

## FOUNDATIONS REVIEW

✓ The respiratory system includes your nose, mouth, throat, lungs, and other structures associated with them. Its main job is to take oxygen gas into your lungs so that it is available for the blood to pick up and carry to the rest of your body.

✓ The air we breathe in takes a long journey through our respiratory system—from the nose to the lungs. Air enters our bodies through our nose and moves from the nasal cavity down through the pharynx. At the bottom of the pharynx, there are two openings, one of which, the larynx (voice box), is connected to the trachea, or windpipe. The trachea is held open by small, C-shaped rings of cartilage so that air can easily move through it. The air then travels down the trachea and into your chest, and the air then divides into two smaller tubes called bronchi that each go to a lung.

✓ Ventilation is the movement of air in and out of your lungs. It consists of two steps, inhalation, or bringing air into the lungs, and exhalation, moving air out of the lungs. The primary muscle controlling ventilation is the diaphragm. When the diaphragm is relaxed, it curves upward. During inhalation, the diaphragm contracts and flattens out, increasing the volume of the chest cavity. The increase in chest cavity volume creates a slight vacuum, bringing a rush of air into the lungs. Exhalation occurs when the diaphragm relaxes, causing the volume of the chest cavity to get smaller and the air to move back out.

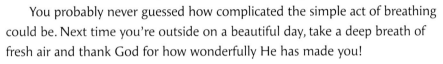

# The Breath of Life

We opened this chapter with a verse taken from the book of Genesis describing how God breathed life into man. The Hebrew word used here is *Ruach*. This word shows up many times in the Old Testament. For example, this is the same word used in the book of Ezekiel when we read about him prophesying to "breathe" life back into the dry bones. But this word is not always translated into the English word "breathe." Sometimes it is translated as "wind" or "spirit." In both of these cases, God is breathing physical life into nonliving forms.

In the New Testament, we start to see something different. After Christ rises from the dead, He appears to His disciples. He shows them the wounds in His hands and feet, and then we read: "Jesus said to them again, 'Peace be with you. As the Father has sent me, even so I send you.' And when he had said this, he breathed on them, and said to them, 'Receive the Holy Spirit'" (Jn 20:21–22). Jesus breathes on His disciples and they receive the Holy Spirit. This is followed fifty days later by the Holy Spirit descending on the Apostles in the Upper Room. "When the day of Pentecost had come, they were all together in one place. And suddenly a sound came from heaven like the rush of a mighty wind, and it filled all the house where they were sitting. And there appeared to them tongues as of fire, distributed and resting on each one of them. And they were all filled with the Holy Spirit and began to speak in other tongues, as the Spirit gave them utterance" (Acts 2:1–4).

The disciples that Christ appeared to after His resurrection and the apostles in the Upper Room were already alive physically. This breath of God, the Holy Spirit, brings *spiritual* life. It gave the disciples the courage they needed to go out into the world and share the gospel with the world. We celebrate this each year during the feast of Pentecost. It is considered the birthday of the Church.

In the sacrament of Confirmation, those of us who have been baptized into the Church receive more fully the gifts of the Holy Spirit so that we can become soldiers for Christ and His Church. The gifts of the Holy Spirit are wisdom, fortitude, understanding, counsel, knowledge, piety, and fear of the Lord. By increasing these gifts in our lives, we are united more closely to God and receive special grace to be witnesses of God.

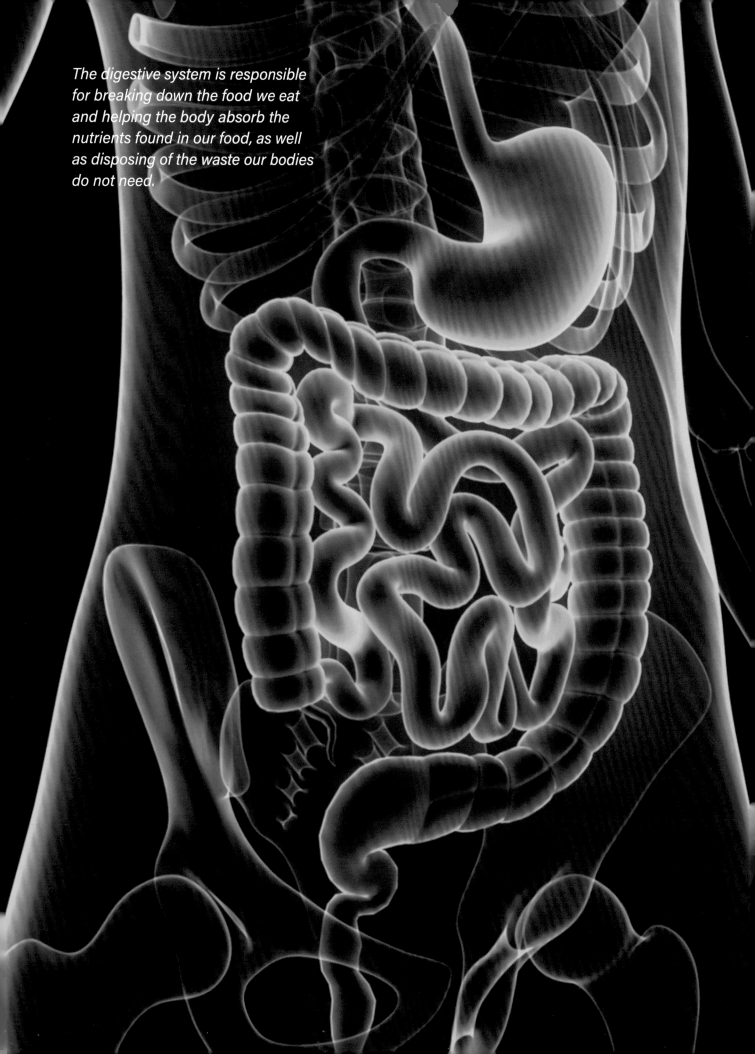

*The digestive system is responsible for breaking down the food we eat and helping the body absorb the nutrients found in our food, as well as disposing of the waste our bodies do not need.*

# CHAPTER 10

## THE DIGESTIVE SYSTEM

## WE ALL HAVE TO EAT!

We all have a favorite food, or two, or three. As a kid, I always asked for lasagna when I was given the chance to choose. But we don't eat just because we like the taste of food. Food provides the necessary energy that we need for our body, organs, and cells to carry out their jobs.

Even so, our cells cannot "eat" a hamburger or a slice of pizza. Our cells need energy in the form of small molecules. So how does the food we eat get broken down into a form that cells can use? And how do those molecules get to all the cells in our body? This is the purpose of the **digestive system**. In this chapter, we will take a tour of the digestive system and learn how it is able to carry out this very important job.

## FUNCTIONS OF THE DIGESTIVE SYSTEM

Before we begin our tour, it is helpful to introduce the main functions of the digestive system. There are five main functions of the digestive system:

1. Mechanical digestion
2. Chemical digestion
3. Absorption
4. Elimination
5. Movement

*Digestive System Fun Fact:*
*Your large intestine is actually shorter than your small intestine if you were to stretch them both out. It receives its name because it has a wider diameter (it is thicker).*

Let's take each of these one at a time.

The digestive system is responsible for the *mechanical digestion* of food. This means that it tears the food into smaller pieces. An example of this is the chewing we do with our teeth.

Another function of the digestive system is *chemical digestion*. In chemical digestion, the body uses chemicals to break food down further (rather than "mechanics" like teeth chewing). This is done using special proteins called **enzymes**. There are thousands of different enzymes in our bodies, each with a particular job. In the case of the digestive system, these enzymes take large molecules and break them apart into their individual subunits. For example, bread is made of large molecules called carbohydrates. Carbohydrates are chains of simple sugar subunits. The enzyme amylase takes the large carbohydrate molecule and breaks it apart into the individual sugar subunits, sort of like how you can break apart a LEGO creation into individual bricks (although this would be more of a mechanical activity than chemical since you are doing it with your hands). The enzymes breaking down the carbohydrates will make it easier for the body to digest the bread and get the energy it needs from the bread.

Once the food is broken down into its individual subunits, the digestive system must absorb those nutrients into the blood. This is called *absorption*. But not everything we eat is nutritious and gets absorbed. Things that are not absorbed must be removed from the digestive system. We call the removal

# DIGESTIVE SYSTEM

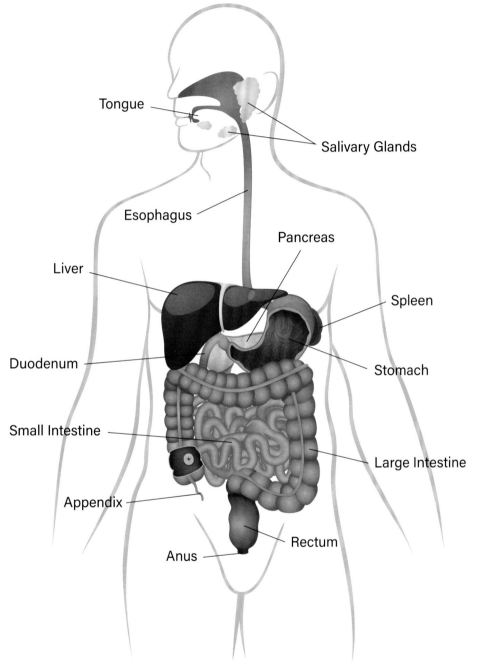

*Just over a dozen body parts carry out the five functions of the digestive system.*

of the waste *elimination*. Finally, in order for all of these functions to work together, the food is constantly being moved down through the digestive tract. The digested contents are moved by contractions of muscular tissue found within the walls of the digestive organs. This, of course, is the final function of *movement*.

Now let's start our journey through the digestive system and see where all these different functions are taking place.

## FIRST STOP: THE MOUTH

The mouth is a doorway into our digestive system. When we put food in our mouth, we call this **ingestion**. The mouth is also where we begin to break down our food. Food is broken down in the mouth through mechanical digestion. As we chew, it breaks down the food into smaller pieces. As we just learned, food is also broken down through chemical digestion. When we put food into our mouths, or if we smell some cookies baking in the oven, it causes our mouth to begin to fill with a watery substance called **saliva**. The saliva has enzymes that break down the food through chemical reactions.

Try this little experiment to see how saliva helps to break down food. Take a small soda cracker or piece of white bread and place it in your mouth. Do not chew it! Rather, allow it to sit on your tongue for a few minutes. You will notice that the food starts to get soft and dissolve. This is because the enzymes in your saliva are chemically breaking down the food.

Mechanical and chemical digestion in the mouth starts the process of breaking down the food into small molecules. But more work is needed. At this point, the chewed food is pushed to the back of the mouth with the tongue and swallowed. When you swallow, a flap of cartilage in your throat called the epiglottis covers the windpipe. This directs the food back towards a muscular tube called the **esophagus** that connects the throat with the stomach. Once the food is in the esophagus, it is pushed down into the stomach. This takes about two to three seconds.

## SECOND STOP: THE STOMACH

The **stomach** is like a storage bag for your food. If you did not have a stomach, you would have to eat constantly because your body needs energy constantly. If you eat a lot of food, the stomach can stretch to hold all that food, which stretches out the time between when you need to eat.

Once the food enters the stomach, it continues to get broken down. Muscular tissue in the stomach walls contract. This causes your stomach to squeeze and churn the food in the stomach, almost like a garbage disposal or

### Why Does My Stomach Growl?

It is not uncommon for your stomach to "growl" or rumble when you are hungry. The noise is caused by the movement of food, fluid, and mucus through your stomach and intestines. When you're hungry, your brain sends a message to the muscular tissue in the walls of the digestive organs. The muscles start contracting, which causes the noise. Grumbling noises can also keep happening after you eat. As the muscles contract, they continue to push and squeeze the digested food through your digestive tract, producing more noise.

blender churns up the food. The churning also mixes the food with stomach juices. The juices are released from small cells lining the inside of the stomach. The juices are very acidic. In fact, the stomach juices are more acidic than soda, coffee, or orange juice! Thankfully, the stomach wall is made to withstand the very acidic environment. There are also special enzymes in the stomach juice that help to break down food even further. All of this mechanical and chemical digestion produces a fluid mixture called **chyme**. The chyme is slowly released out of the stomach into the small intestine. Only about one to two teaspoons of chyme is released at a time. As a result, it can take three to five hours for your stomach to empty completely.

## THIRD STOP: SMALL INTESTINE

The **small intestine** is the place in your digestive system where all the nutrients that have been broken down get absorbed into your body. Nutrients such as sugars and the subunits of proteins (these are called amino acids) get carried across the wall of the small intestine into the capillaries. The blood then carries the nutrients throughout the entire body to the cells that need them.

SMALL INTESTINE

In order to absorb large amounts of nutrients, the small intestine has a very special structure. First, it is very long. If you were to stretch out the entire length of an adult's small intestine, it would stretch about twenty feet! So why do we call it small? Well, because it has a smaller diameter (about an inch) compared to the large intestine (which we will speak about shortly). The long length of the small intestine means that the digested food spends more time in it and there is more time for the nutrients to get absorbed into the blood. Second, the inside of the small intestine is lined with small fingerlike structures called **villi** that provide more area for absorption.

Here is a simple way to think about why the villi are important. Imagine if you drew a straight line from point A to point B. How long is the line? Now, what if I draw a wavy line between point A and point B. Now how long is the line? If you were to stretch it out, you would see that the wavy line is longer.

This is similar to the villi in the small intestine. They provide more area inside the intestines for nutrients to be absorbed just like the wavy line. If the inside of the small intestine was smooth, it would be like the straight line.

When the chyme first enters the small intestine, other chemicals and enzymes also enter. These come from the liver and the pancreas. The **liver** is a large organ found on the right side of the abdomen just below your ribs. It has many functions, including removing toxins from the body. The liver also produces a special chemical called **bile**. Bile helps break down fats in your food. The bile is stored in a little sac called the **gallbladder** found underneath the

# LIVER ANATOMY

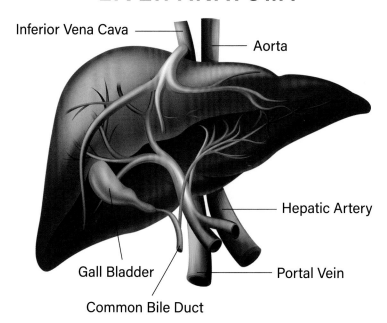

liver. When digested food enters the small intestine, bile from the gallbladder gets mixed in to break down the fats. The **pancreas** is found on the left side of your abdominal region. It releases several different enzymes that help to break down sugars and proteins. These are also mixed with the digested food at the beginning of the small intestine. The chyme continues its path through the small intestine and, finally, the small intestine joins with the large intestine. This takes about three hours.

## FOURTH STOP: LARGE INTESTINE

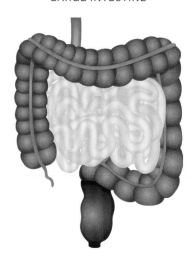

LARGE INTESTINE

The small intestine eventually joins with the **large intestine** in the bottom right side of the abdomen. It is called the large intestine because it has a larger diameter than the small intestine, about 2.5 inches. However, it is shorter than the small intestine, reaching only five feet in length. The **appendix** is a small fingerlike tail off the end of the large intestine near where it attaches to the small intestine. For many years, scientists thought the appendix had no purpose. Now we know that it provides a protective home for good bacteria that live inside our intestines. Yes, you have bacteria living inside you! These bacteria help to break down some of the food products that our bodies can't digest on their own. The presence of these bacteria also helps keep out bad bacteria that might make you sick.

The contents in the large intestine do not contain many nutrients. It is mostly indigestible material such as plant fiber. As the waste material moves through the large intestine, water is absorbed. This causes the waste matter to thicken into a solid mass. Biologists refer to the waste as feces. This gets stored in the rectum at the end of the large intestine. The feces then exits the rectum through an opening called the anus in a process called defecation.

## Blood Sugar and Diabetes

The pancreas has another important role. It helps the body keep the right amount of sugar (glucose) in your blood. Glucose is used by our cells to make energy. If we do not have enough glucose, then we begin to feel tired. If it drops too low, it can cause a person to feel dizzy. Too much glucose in the blood can also cause serious medical problems. The pancreas makes a special protein called insulin that helps the body keep the right levels of glucose in our blood.

**Diabetes** is a disease when the body cannot keep the blood glucose levels steady. Someone with diabetes has to be careful about what they eat and when they eat it. For example, they may not be able to enjoy a large piece of chocolate cake with a scoop of ice cream on the side because it has too much sugar. Their body does not know what to do with it all. Thankfully, there are medicines available to people with diabetes that help their bodies control the amount of glucose in the blood.

Our journey through the digestive system is complete. As you can see, like with so many of the other systems we have studied, each digestive organ has just the right structure for it to carry out its specific function. It is truly amazing all the steps that take place for our body to break down the food we eat to give us the nutrients we need to live and grow.

### FOUNDATIONS REVIEW

✓ Our cells need energy in the form of small molecules. The purpose of the digestive system is to break down our food and get those molecules to the cells. It does this through its five primary functions: (1) mechanical digestion, (2) chemical digestion, (3) absorption, (4) elimination, and (5) movement.

✓ Food first enters through the mouth, where it is broken down both mechanically (by the teeth) and chemically (by saliva). It then goes down the esophagus and into the stomach, where it is further broken down by the stomach contracting to churn up the food as well as through acidic juices released from the lining of the stomach wall.

✓ After leaving the stomach, the food goes into the small and large intestines, where nutrients get absorbed and chemicals from the liver and pancreas (bile and enzymes) further break down the food. Once it leaves the intestines, the waste material (feces) gets stored in the rectum at the end of the large intestine. The feces then exit the rectum through an opening called the anus in a process called defecation.

# The Eucharist: Spiritual Food for Our Souls

In this last chapter, we learned a lot about how our body is able to take in food, break it down, and absorb the necessary nutrients that we need to grow and be healthy. As we eat physical food, our bodies break it down into smaller components and it becomes part of our physical bodies. This is because we, as living human beings, are greater than the food we eat. In a similar way, our spiritual bodies also need food to nourish our souls and help us in our spiritual growth.

One way to nourish our spiritual growth is to spend time every day in prayer. When we pray, we are communicating with God. This helps us strengthen our relationship with Him. There are many saints who have written about the life of prayer. One of these is Saint Teresa of Ávila. She spent many years struggling with her prayer life because she was often distracted by the events and things of the world. Eventually, she recognized that she needed to focus on *who* she was speaking with rather than simply saying the words. If we work to focus on the truth that when we pray, we are talking with God and asking Him for our needs, it will help us grow in our spiritual life. Saint Teresa of Ávila is considered one of the Doctors of the Church because of the writings she left us about how to grow in our spiritual lives.

Perhaps the most obvious source of spiritual food is the Eucharist. "Jesus said to them, 'I am the bread of life;

he who comes to me shall not hunger, and he who believes in me shall never thirst. . . . I am the living bread which came down from heaven; if any one eats of this bread, he will live forever; and the bread which I shall give for the life of the world is my flesh'" (Jn 6:35, 51). What an amazing gift God gives us in the Eucharist! Every time we receive Jesus in Holy Communion, our souls are strengthened and nourished.

It is through Holy Communion that the grace we received in baptism is preserved, renewed, and increased (see *CCC* 1392). These graces help us fight against temptations and against mortal sin in our lives. Saint Ignatius of Loyola said, "One of the most admirable effects of Holy Communion is to preserve the soul from sin, and to help those who fall through weakness to rise again."

Reception of Holy Communion also increases our love for God and others. This is why the Church teaches that the Eucharist is the source and summit of the Christian Faith. Without it, we would be spiritually malnourished, lacking the food we need to grow stronger in our Catholic Faith. But unlike physical food that becomes part of our bodies, the spiritual food we receive in Holy Communion *changes us*. When we receive the Body of Christ in a proper manner, we become more like Him. And we also become part of the mystical Body of Christ. Knowing this, we should follow the example of the saints who have gone before us and go often to receive Our Lord in Holy Communion.

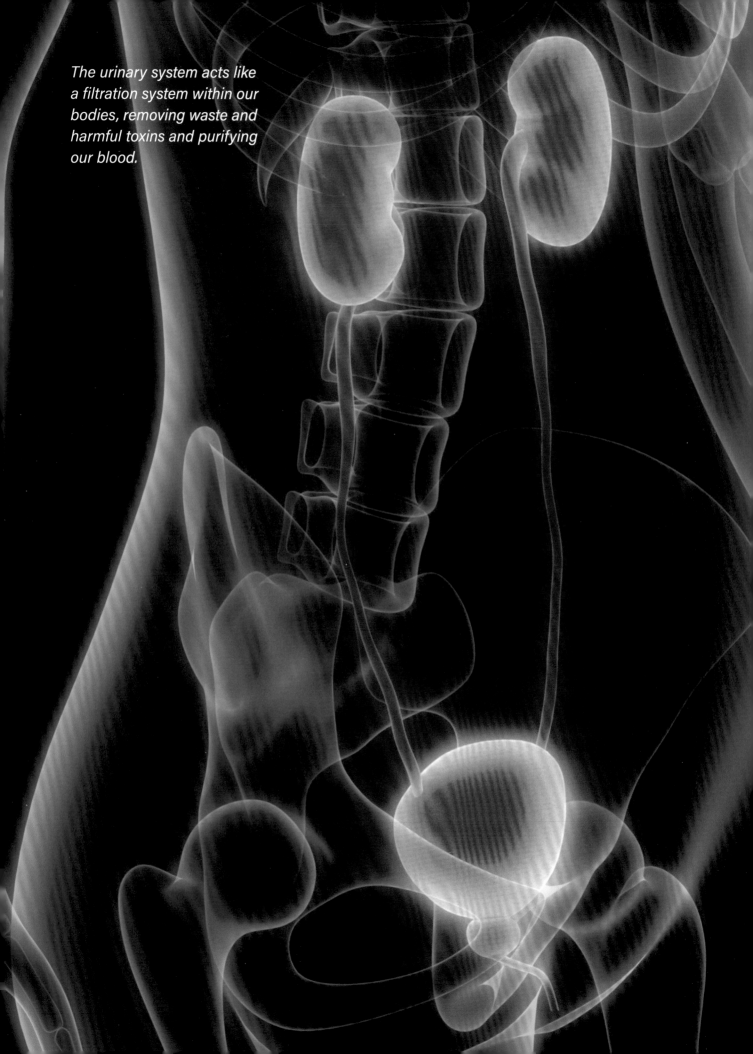

*The urinary system acts like a filtration system within our bodies, removing waste and harmful toxins and purifying our blood.*

# CHAPTER 11

*THE URINARY SYSTEM*

*Remember:*

*It is important to take care of your kidneys by drinking plenty of water. This keeps the main organ responsible for filtering toxins and waste from your body clean.*

### THE BODY'S NATURAL FILTER

In our last chapter, we learned about the digestive system and its role in providing nutrients for our cells to use. The cells break down nutrients such as glucose and proteins to make energy that they use to carry out their normal functions. But during this process, the cells also produce waste products. If these waste products were to build up in our cells and in our body, it would cause a lot of problems and make us feel sick.

This is where the kidneys come into play. The **kidneys** are an organ that act as a special filter to remove the waste products from our blood, making it clean and ready to use. You could think of them like a water filtration system removing the waste from our bodies. (When you get water from your refrigerator, there is a filter up inside it that cleans the water before it comes to your cup.) The kidneys work as part of the **urinary system** to remove waste from our bodies (through urine) and filter our blood. In this chapter, we will explore how the kidneys and the urinary system works.

## ANATOMY OF THE KIDNEY

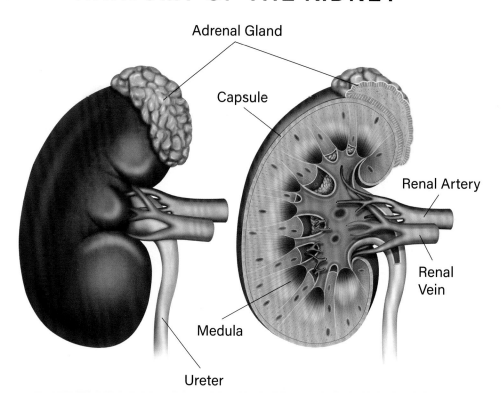

### THE KIDNEYS: WHAT ARE THEY?

The kidneys are bean-shaped organs found along either side of your backbone in your lower back (there are two of them). They sit about the same height as your elbows and are about the size of your fist. They hide just below the bottom ribs and are surrounded by a lot of fat; the ribs and the fat protect the kidneys from being damaged if you were to get hit hard on your back.

These special organs have several different functions. One role of the kidneys is to release a hormone called erythropoietin. A **hormone** is a special protein that carries a message from one organ to another organ. The hormone **erythropoietin** is made in the kidneys and carries a message to the bones telling them to make more red blood cells. Another function of the kidneys is to make the active form of vitamin D. We need vitamin D to help us absorb calcium from our food. While both of these are important functions of the kidneys, the primary job of the kidneys is to help to filter out and remove waste from the blood. Let's take a closer look inside a kidney and learn how it works.

## FILTERING BLOOD

First, the blood enters the kidneys through a big artery called the **renal artery**. Once inside the kidneys, the renal artery branches into smaller and smaller blood vessels. Some of these tiny blood vessels come together into a small ball of capillaries called the **glomerulus**. The glomerulus is enclosed in a cup-shaped capsule. After the blood moves through the glomerulus, it leaves the ball of capillaries through another small blood vessel. These come together to form larger and larger veins until at last they exit the kidney as the renal vein.

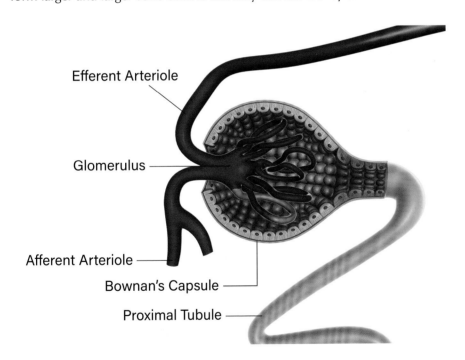

*You have probably never heard of the glomerulus, but it is a vitally important part of your urinary system. It is the place deep within the kidneys where the filtration of our blood takes place.*

Filtration of the blood takes place in the glomerulus. As the blood moves through the glomerulus, water, sugars, salts, and a waste product called urea are pushed out of the blood through small slits in the walls of the capillaries. The filtered components, called the filtrate, collect in the capsule surrounding the glomerulus and then move into a set of tiny tubules. These tiny tubes are called **nephrons**. There are millions of nephrons found inside each kidney. As the filtrate moves through the tubules of the nephrons, the blood in the kidneys reabsorbs most of the water, salts, nutrients, and any other necessary components. If there is an excess of certain molecules in the blood, these get removed or secreted out of the blood and moved into the filtrate. The filtrate from the nephrons collects in the center part of the kidney, called the renal pelvis. Once in the renal pelvis, the filtrate is referred to as urine. Urine empties out of the kidneys through a tube called the **ureter**, where it is carried to the bladder.

## THE BLADDER: A STORAGE BAG

The **bladder** works kind of like a balloon. The walls of the bladder have stretchy elastic tissue and a thick layer of muscular tissue. As urine from the ureters constantly fills the bladder, it stretches and gets bigger. The bladder is held closed by a small round muscle at the bottom called a sphincter muscle. This is kind of like taking a thick rubber band and wrapping it around the opening of the bladder. When the bladder gets very full, the walls of the bladder stretch; this is what lets you know you need to use the bathroom, that uncomfortable feeling of pressure. The stretching activates special nerve cells in the bladder walls. These send a message to the nervous system. In response, the thick layer of muscle tissue in the walls of the bladder contract and the

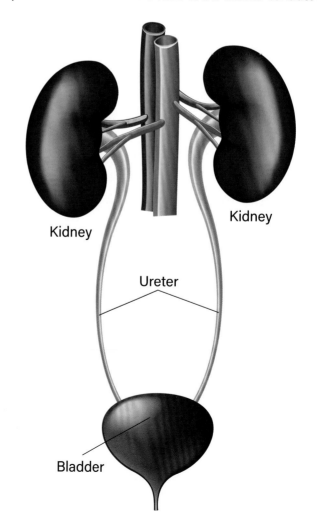

*Urinary System Fun Fact:*

*A properly working kidney can filter about a half a cup of blood every minute. This quickly adds up to about forty-five gallons of blood in the course of one day!*

## Kidney Stones

Kidney stones are small crystal structures made of salt or uric acid that form in the kidneys. It is unclear exactly why some people develop kidney stones. Some things that could cause them may be not drinking enough water on a regular basis, eating foods high in salt, or it could be a side effect of another medical condition. If a kidney stone forms in the kidney, it has to come out with the urine. If the kidney stones are small, this may not be too big of a problem. However, sometimes they can grow to be rather large. When this happens, passing a kidney stone can be a rather painful process. Sometimes doctors will use sound waves to try to break up the kidney stone into smaller pieces, making it easier to pass.

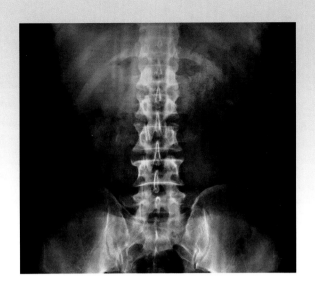

sphincter muscle around the opening relaxes. This lets the urine exit out of the bladder through a tube called the **urethra**. Once all the urine has been released out of the bladder, it returns back to a small flat bag again and starts to fill with urine once more.

While you may think of urine as being rather gross, it is actually very clean. In fact, urine is sterile. This means that it does not contain any germs or bacteria. Urine is made mostly of water. Dissolved in the water are chemicals such as salts and urea. It is normally yellow in color and clear. If you drink a lot of water during the day, then there is more than enough water in the blood. The kidneys remove the extra water, and it ends up in the urine. This causes your kidneys to make a lot of urine that will have a lighter yellow color. If you do not drink enough water during the day, your kidneys will reabsorb more water out of the filtrate. This causes the urine to become very concentrated. There will be less urine and it will have a much darker color. If this is the case, it means you should probably drink more water. If urine has a cloudy appearance, this could be a sign of a bacterial infection. These can easily be treated by antibiotics prescribed by a doctor.

## HEALTHY KIDNEYS

The kidneys are very important to our body's ability to work properly. If the kidneys stop working, our body would quickly fill up with waste and toxins and we would feel very sick. This is why we want our kidneys to stay healthy. The best way to keep your kidneys working properly is to live a healthy life. We have talked about many ways to do this in previous chapters. Eating healthy foods and staying active are both easy ways to keep our bodies strong and healthy. Drinking plenty

*Urinary System Fun Fact:*

*The human bladder can stretch to hold about 400ml of urine—that's almost two full cups!*

of water is also good for our kidneys. There are also things we can avoid to keep our kidneys healthy. For example, too much alcohol, regular smoking, or taking too much aspirin can cause damage to the kidneys.

As people get older and if they have other medical problems, their kidneys might start to have a hard time filtering all that blood. If that starts to happen, a doctor may recommend that his patient receive kidney dialysis. This is a medical procedure that works to remove the waste, salt, and extra water in the blood. In short, it is a machine that does the job of the kidneys. A person who has kidney failure usually must go to a special clinic three times a week to receive a dialysis treatment.

Unfortunately, once kidneys stop working, they do not start again. Another option is for someone to receive a kidney transplant. Amazingly, God made us so that we can live with just one half of one kidney. In very special cases, doctors can take a working kidney from one person and put it into the patient with the non-working kidney. This takes a lot of careful work from specially trained doctors, but it gives the patient a new and working kidney again. Thank God for smart doctors!

With the urinary system behind us, we can move on now to our last chapter.

## FOUNDATIONS REVIEW

✓ The kidneys are an organ that act as a special filter to remove the waste products from our blood, making it clean and ready to use. The kidneys work as part of the urinary system to remove waste from our bodies.

✓ The two main organs that comprise the urinary system are the kidneys (we have two of them) and the bladder. The kidneys are bean-shaped organs found along either side of your backbone in your lower back, while your bladder is a balloon-like sac that can stretch when filled with urine and eventually passes it out through a tube called the urethra.

✓ The kidneys are very important to our body's ability to work properly. If the kidneys stop working, our body would quickly fill up with waste and toxins and we would feel very sick. The best way to keep your kidneys working properly is to live a healthy life and to drink lots of water. There are also things we can avoid to keep our kidneys healthy. For example, too much alcohol, regular smoking, or taking too much aspirin can cause damage to the kidneys.

# Confession: Removing Toxins and Waste from Our Souls

Our kidneys are constantly filtering out the waste products that build up in our blood so that the blood is "clean" and can be used to carry the nutrients and oxygen to all our cells and organs.

As we live our daily lives, we also build up *spiritual* waste in our souls. Because of the original sin of Adam and Eve, each of us is born with the tendency to sin. If we don't regularly clean the sin out of our lives, it can become toxic to our souls. Thankfully, God, through His Church, has given us a way to clean out our souls from the waste of sin. This is through the Sacrament of Penance, or Confession.

In confession, the priest acts in the person of Jesus Christ to forgive each and every one of our sins, so that our souls are made clean again. The Church encourages us to go to confession about once a month because she knows that we need to clean the waste out of our souls often. At the very least, we are required to go to confession once a year.

But confession does more than simply clean our souls. Going to confession often helps us grow in our spiritual lives as well. The grace we receive in confession can give us strength to resist sin. We get better at avoiding sinful behavior because we are working to follow God's will. It helps us to break our bad habits and develop new ones. And, over time, frequent confession helps us to become saintlier.

### Act of Contrition

*O my God, I am heartily sorry for having offended Thee, and I detest all my sins because of Thy just punishments, but most of all because they offend Thee, my God, who are all good and deserving of all my love. I firmly resolve with the help of Thy grace to sin no more and to avoid the near occasion of sin.*

*Amen.*

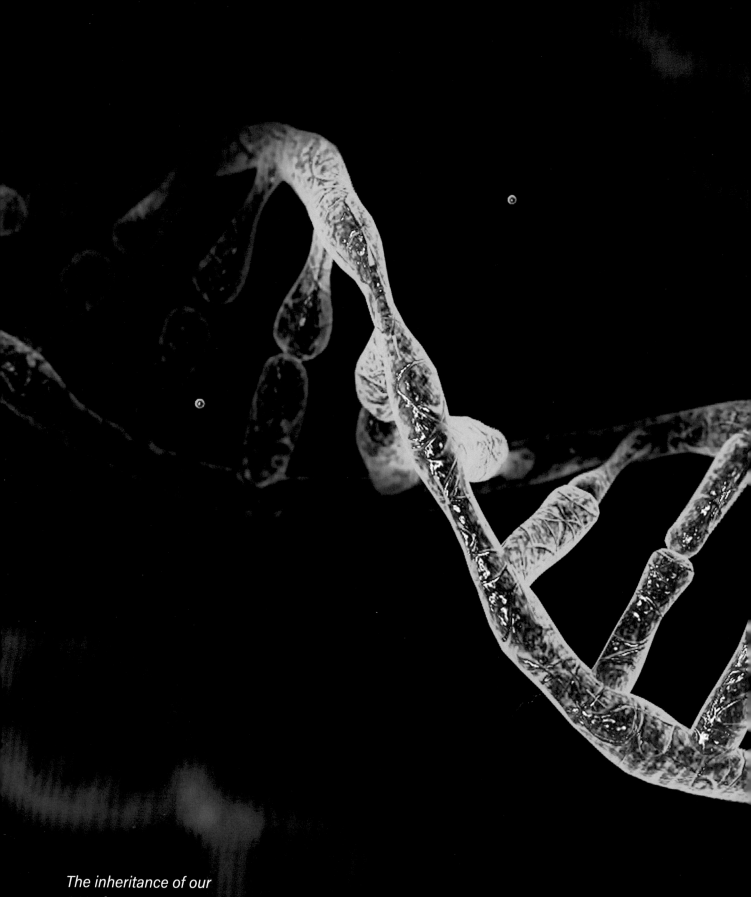

*The inheritance of our DNA from our parents and grandparents is the reason we resemble them.*

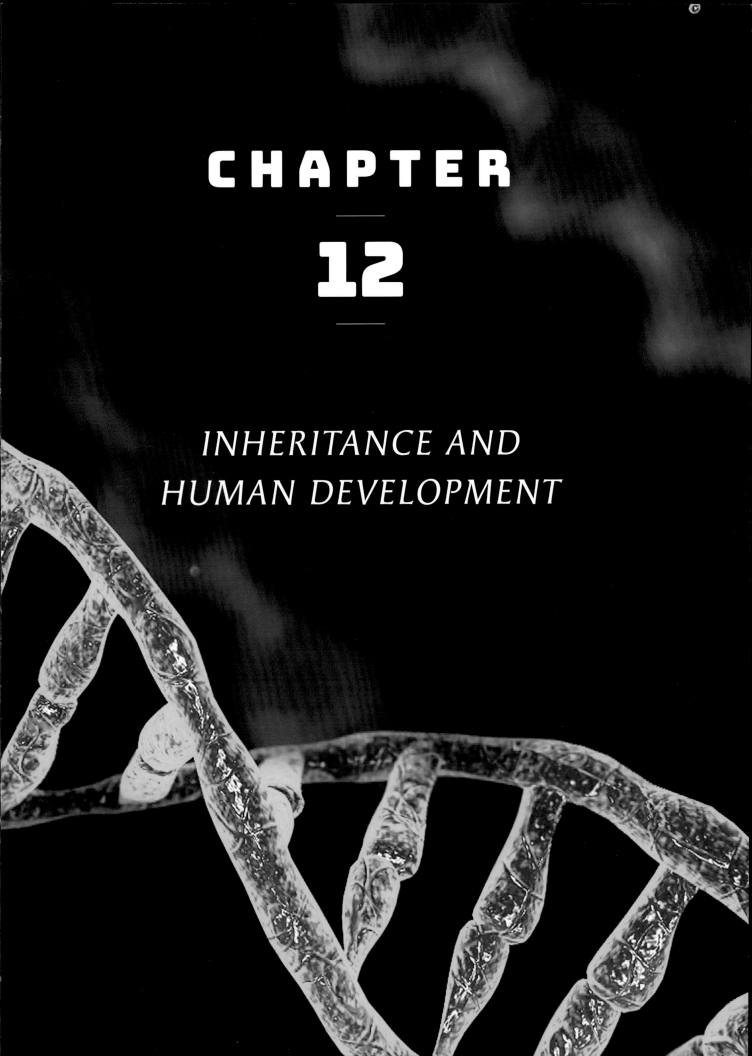

## RETURNING TO THE CELL

Over the past several chapters, we have been learning about all the incredible organ systems found in the human body. These systems work together to give our bodies structure, help us move, allow us to sense things in our world, get oxygen and food for our cells to make energy, and remove waste.

But how do our bodies know how to do all these amazing things? That's a good question! The answer is found deep within each and every one of our cells. So just as we started with the basic unit of a cell and worked our way out, let's go full circle and return to our individual cells to learn how they do what they do.

## DNA: INSTRUCTIONS FOR LIFE

We started this journey by learning about the basic unit of all living things, the cell. Inside the cell, enclosed in the nucleus, is the DNA. DNA is a complex molecule that contains all the "instructions" needed for a cell to do its special job. You can think of it kind of like the code for a computer program. Codes for a computer help it know what to do so that when you click a certain button, there is a certain reaction. The code acts like directions, or instructions, for the computer to follow. But in order for our cells to do the job they were made to do, the information in the "program" must be read. In other words, what good are instructions if they can't be read and understood? In biology, we call this process **gene expression**.

*Gene expression is a two-step process (transcription, translation) by which a DNA molecule is "read" so that a cell knows how to carry out its proper function. Ultimately, it is like the deciphering of a code that makes us who we are and ensures our bodies function as they should.*

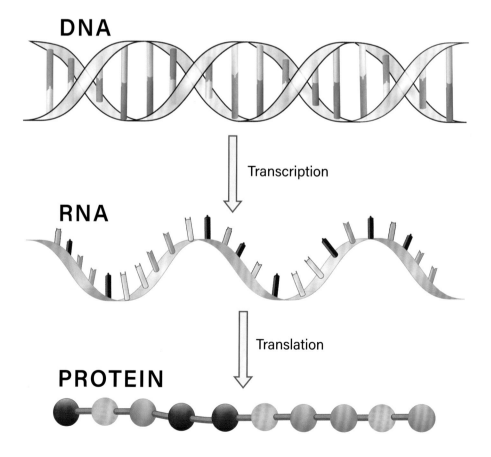

Gene expression is a two-step process that takes the information stored in a segment, or **gene**, of the DNA and uses it to make a particular protein. The first step is called *transcription*. Transcription takes place in the nucleus of the cell where the DNA is located. During transcription, special enzymes make a copy of the gene by making a new molecule called RNA (ribonucleic acid). RNA has the same basic structure as DNA, but it does not store the information. It simply carries the information from inside the nucleus out into the cytoplasm of the cell. We often refer to it as messenger RNA because it is carrying the message from one part of the cell to another. Once its job is done, the messenger RNA is destroyed.

The second step of gene expression is called *translation*. During translation, the RNA is "read" by ribosomes in the cytoplasm. The ribosomes make proteins by stringing together the building blocks or subunits of the protein called amino acids. You can think of this kind of like using letters in the alphabet to make words or sentences. The amino acids are the letters, and they must be put together in the right order to make a protein that works correctly, just as if you want to spell a certain word, you have to put the letters in the right order. If there is a mistake in the order, the protein will not function properly. We say that the proteins are "folded" into their proper shape so that they can be functional.

If this seems a little confusing, let's look at an example to help clarify it.

The enzyme lactase is a protein that breaks down the sugar lactose in milk so that our body can digest it. This protein is made with instructions from a gene found on chromosome 2 in our cells. During transcription, the lactase gene is used to make a messenger RNA molecule. This carries the information to the ribosomes. They read the information on the messenger RNA and build the lactase enzyme. If there is a mistake, or mutation, in the gene for lactase, the amino acids are put together incorrectly, and the enzyme does not work—the lactose (in the milk) does not get broken down properly. This is what may cause someone to be "lactose intolerant." If someone is lactose intolerant, it is probably because they received a bad copy of the gene from their parents, and so dairy products can make them sick (they struggle to digest them).

## INHERITANCE

Have you ever wondered why you look a lot like your mom or dad? This is because you received some of their DNA when you were first formed. **Inheritance** is the passing of traits from parents to children. Each of our cells contains two copies of every chromosome. One of these sets of chromosomes you receive from your dad, and the other you receive from your mom. Together, they make you who you are. This means that you receive or inherit two copies of every gene. Sometimes these genes may be the same. For example, maybe you inherited dark hair from both your mom and your dad. But other times, the two copies may be a little different. When you inherit two different versions of a trait, the *dominant* trait will be the one that you see. If you inherit dark hair from your mom and blonde hair from your dad, the trait

*Remember:*
*Each of our cells contains two copies of every chromosome. One of these sets of chromosomes you receive from your dad, and the other you receive from your mom. Together, they make you who you are!*

for dark hair is dominant over blonde hair. This means that you will not have blonde hair even though you still have a copy of the blonde hair gene. In this case, the blonde hair gene is the *recessive* trait. In order for someone to display or express the recessive trait, you must inherit two copies, one from your mom and one from your dad.

Below are some examples of inherited traits and if they are dominant or recessive. What traits do you have? What traits do your parents and siblings have? You may be able to determine if you inherited the trait from your mom or your dad.

| Dominant | Recessive |
|---|---|
| 1-D Freckles | 1-R No Freckles |
| 2-D Widow's peak | 2-R Straight hairline |
| 3-D Free ear lobes | 3-R Attached ear lobes |
| 4-D Dimples | 4-R No dimples |
| 5-D Bent little finger | 5-R Straight little finger |
| 6-D Tongue roller | 6-R Non-tongue roller |
| 7-D Right-handed | 7-R Left-handed |

## Unusual Patterns of Inheritance

Most inherited traits do not follow a simple inheritance pattern of dominant or recessive. Many traits we think of, such as height, eye color, or skin color are controlled by many genes. We call these polygenic traits (*poly* means many, and *genic* refers to gene). Because many genes contribute to the trait, we observe a whole range of traits. For example, grown adults are not one of two heights (short or tall). They can be 5' 1" or 5' 11" or 6' 8", or any other height.

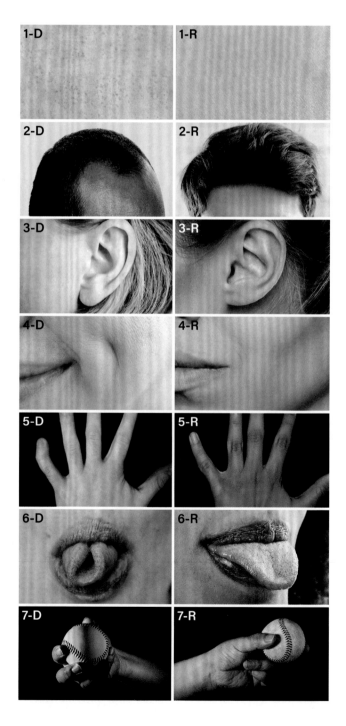

# HUMAN DEVELOPMENT

When a new human life begins as a single cell, it contains all the DNA that is needed to grow into an adult. The first nine months of a human's life is spent inside its mother as it grows and develops all the different organs and organ systems that you have been learning about in the past few chapters. These nine months are broken into two parts. The first part lasts about two months. This is the **embryonic stage**. The second part lasts about seven months. We call that the **fetal stage**.

The embryonic stage begins with a single cell that divides many times and very quickly; this is the start of the creation of the child. To help us see the humanity of this child, let's say she is a little girl, and her name is Lily.

After four days, Lily forms into a small hollow ball made up of hundreds of cells that are all exactly the same. At this point, she is only about the size of a poppyseed! Over the next week, the cells continue to divide, but they also start to change into different types of cells. Some cells will become muscle, bone, or blood; others will develop into the skin or nervous tissue, and others will become parts of the digestive system. But not all these cells become parts of Lily's body—some will form the **placenta**, which is a "temporary" organ that develops in the mother's uterus during pregnancy to provide oxygen and nutrients to the growing baby. The placenta attaches to the wall of the uterus and the baby's umbilical cord is connected to it (you have probably heard before that your umbilical cord is what gives you your belly button).

Lily's first organ to form is the brain at about eighteen days. By day twenty-two, her heart has formed and pumps blood throughout her body. After five weeks, the digestive organs are all starting to form, and by twelve weeks, the digestive system will be complete. Distinct bones, such as those in the arm, are visible at six weeks. The bones begin to connect to muscle in the following week, allowing her to move and kick. By this time, Lily has fully formed skin, hair, and fingernails and is about the size of a grape. At eight weeks, her kidneys are working to filter blood and make urine. The respiratory system begins to form around four weeks, and by six months, her lungs have formed. While most of the organs and organ systems are in place by the end of the embryonic period, Lily is still very small, only about an inch from her head to her bottom.

*Human Development Fun Fact:*
*The oldest person recorded to live in modern times was a French woman named Jeanne Calment who lived to be 124 years old. She died in August of 1997. Of course, if you go back to Old Testament times, we find many people who lived much longer. According to the book of Genesis, Methuselah lived to be 969 years old (see Gn 5:27)!*

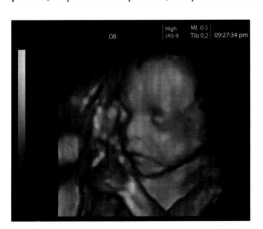

She still has a lot of growing to do before she is ready to be born.

In the fetal stage, growth happens very quickly. By three months, she will be three inches, and by six months, she will have grown to over eight inches from head to bottom. Just before Lily is born, she measures about twelve inches from head to bottom and may weigh around eight pounds.

# HUMAN DEVELOPMENT IN THE WOMB

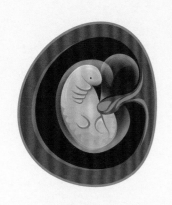

1 MONTH

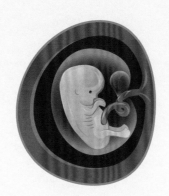

2 MONTHS

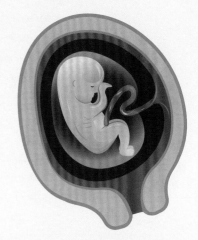

3 MONTHS

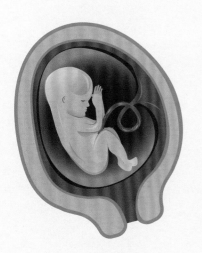

4 MONTHS

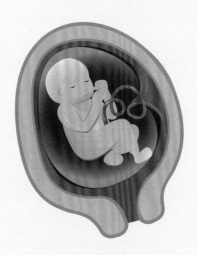

5 MONTHS

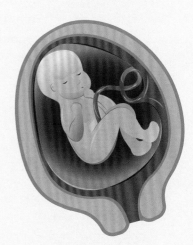

6 MONTHS

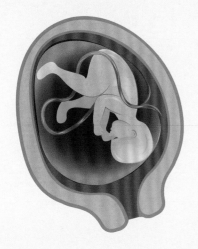

7 MONTHS

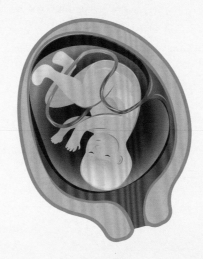

8 MONTHS

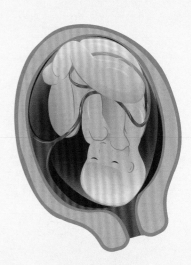

9 MONTHS

It takes about nine months for a baby like Lily to be big enough to live outside of her mother.

After birth, Lily will of course keep growing. If you have younger brothers or sisters, you may have been able to watch them grow from helpless infants to toddlers who crawl and walk, to four-year-olds who talk, run, and jump. You will keep on growing until you get to be around twenty years old, and then most people stop getting taller. But they still have a lot of life left to live. A healthy person can live well past eighty years old, and some people live to be one hundred! It's important to continue to take good care of our bodies so that we can be healthy and strong. This will allow us to live our lives in accordance with God's will, carrying out our vocations to the best of our abilities.

Well, this concludes our twelve chapters, but before you close this book, make sure to flip ahead and read the conclusion for some final thoughts.

## FOUNDATIONS REVIEW

✓ In order for the systems in our bodies to function, the cells that make them up must be able to read the DNA instructions found inside the nucleus. This process is known as gene expression and has two steps: (1) *Transcription* takes place in the nucleus of the cell when special enzymes make a copy of the gene by making a new molecule called RNA (ribonucleic acid), which is like a messenger that carries the information from inside the nucleus out into the cytoplasm of the cell. (2) *Translation* is the process by which the RNA is "read" by ribosomes in the cytoplasm. They do this by stringing together the building blocks or subunits of the protein called amino acids. They must be put together in the right order to make a protein that works correctly.

✓ Inheritance is the passing of traits from parents to children. One set of chromosomes we receive from our dad, the other from our mom. Certain traits in the gene can be dominant, while others are recessive. The dominant trait will win out. Only if you receive two of the recessive genes can you display that trait (for example, you must receive blonde hair from your mom and your dad to have blonde hair).

✓ When a new human life begins as a single cell, it contains all the DNA that is needed to grow into an adult. The first nine months of a human's life is spent inside its mother as it grows and develops all the different organs and organ systems. These nine months are broken into two parts. The first part lasts about two months. This is the embryonic stage. The second part lasts about seven months. We call that the fetal stage.

# Conclusion

We have completed our overview of the human body. Throughout the course of this book, we learned about the structure of cells and some of their basic functions, and we read about how cells come together to make many different types of tissues that are the basic material for all of our organs and systems. I hope you have enjoyed learning about the cells and systems of the human body. If you did enjoy it, there is much more to learn about how all of these systems work. Believe it or not, we only covered the basics here.

Isn't it amazing how intricately are bodies are made? I continue to be blown away by how each cell and each organ has the exact structure they need to carry out their specific functions. In addition, all of our organs and systems work together so that we can live every day. Life is an incredible gift given to us by God. For this we can join with the Psalmist in praising God for creating us: "For you formed my inward parts; you knitted me together in my mother's womb. I praise you, for I am wondrously made. Wonderful are your works! You know me right well" (Ps 139:13–14).

Life is a gift that we should not take for granted. We need to be good stewards of our body. This means we need to take care of ourselves by not causing harm to our body and living a healthy lifestyle. We have mentioned several ways to do this throughout our book already. Some of these include eating a well-balanced diet, getting enough sleep, staying active, and avoiding things that could cause your body harm, such as smoking. I hope that as you grow older, you will continue to take good care of your body so that you can give glory to God in all you do.

# AMAZING FACTS ABOUT CELLS & SYSTEMS

- An English scientist by the name of Robert Hooke was the first to come up with the name "cells." He studied things through a microscope and drew sketches of them. One he drew was of dead plant material, called cork. Hooke cut a thin slice of a piece of cork and looked at it under his microscope. He saw that the cork looked like little boxes, or pores, like what you might find in a honeycomb. Hooke called these boxes cells.

- One of the things we find in every cell is genetic material in the form of DNA (Deoxyribonucleic acid). This genetic material contains the "instructions" that tell the cell what organelles to build and how to carry out its job. DNA is like the directions that make you who you are (how tall you are, the color of your skin, etc.), like blueprints for how to build a house.

- Most cells are too small to be seen without a microscope, but there are some cells that are bigger than others. For example, an egg is a single cell. The biggest egg is laid by the ostrich. It is about six inches long and weighs about three pounds!

- Nerve cells make up our brain and other nervous structures throughout our body. Their job is to carry messages from one part of our body to another. We have nerves stretching from our spine to the bottom of our toe. Giraffes have even longer nerve cells. They have nerves that run the entire length of their long necks, between six and eight feet long! However, the longest nerve cell is found in the giant squid. These sea animals can be up to forty-three feet long and have nerves up to thirty-nine feet long.

- The phospholipid molecule is very special because it is both hydrophilic (loves water) and hydrophobic (fears water). If you were to take thousands of phospholipid molecules and put them into a cup of water, they would arrange themselves into two rows. Biologists call this a phospholipid bilayer because there are two layers of phospholipids. The heads would be on the top and bottom, close to the water. In the middle would be the long hydrophobic tails hiding away from the water.

- Human cells have twenty-three pairs, or forty-six chromosomes. If you stretched out a single chromosome, it would be about six feet long!

- Plant cells have a cell wall and animal cells do not. Another difference between plant and animal cells is that plants have a special organelle called a chloroplast, which is a small organelle that captures light energy from the sun. It uses the light energy to transform carbon dioxide gas and water into chemical energy through the process of photosynthesis. This is how plants make their own food. You can think of the chloroplast kind of like a solar panel. It collects light from the sun and transforms it into energy that the plant can use.

- The ability to reproduce is a key characteristic to all living things. If living things could not reproduce, then life on Earth would not be able to continue to exist.

*Cells & Systems Fun Fact:*
*Our skin is the largest organ in our bodies.*

- Most human cells divide once every day, but some cells divide faster than others. For example, your skin cells take about one hour to divide. Nerve cells, on the other hand, don't divide. They form when a baby is growing in its mother's belly and those nerve cells stay with you for your entire life.

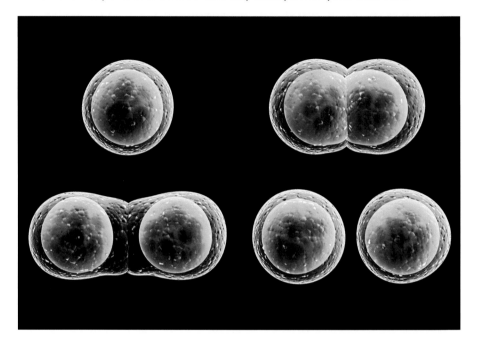

- Our cells will not divide if they are damaged or if there is something wrong with them. If something is wrong, the cell will try to fix the problem so it does not pass the problem on to the new daughter cells. For example, cells sometimes make a mistake when copying the DNA. When this happens, the cell will try to fix the mistake in the DNA. If the problem cannot be fixed, then the cell will self-destruct. This happens so that the damaged cell does not make more damaged cells (it doesn't copy itself and thus copy the mistake).

- Our skin is the largest organ on our bodies.

- Some estimates say that a person loses thirty thousand dead skin cells every day! Those cells that fall off are replaced. The very bottom layers of cells in the epidermis are constantly dividing to replace the cells that fell off. In fact, you get a "new" suit of skin about once a month.

- The blood vessels in our skin help keep our body temperature steady. When we get too warm, more blood runs through the blood vessels in our skin. This releases heat from our bodies and helps us cool off. Another way our skin helps us to cool off is that it contains tiny little sacs called glands. These glands produce sweat when we get hot. The sweat evaporates and helps cool us off.

- When you get cold, hairs on your body tend to stand up to try to help you get warm.

- The color of our hair and skin comes from a special pigment called melanin that is found deep in the epidermis. People have different amounts of melanin in their skin. Someone with a lot of melanin will have darker skin.

*Cells & Systems Fun Fact:*
*The hairs in our nose act like a filter to keep out dust and pollen and other particles that would otherwise make it into our respiratory system.*

Someone with very little melanin in his skin has a pinker appearance because the color of the blood shows through his skin. Freckles form when there is a lot of melanin that collects in one place in the skin. These are easier to see on people with paler skin.

- If you were to look at bone under a microscope, you would see that the minerals are laid down in a pattern of circles. It looks a lot like a tree stump that is left behind after cutting down a tree.

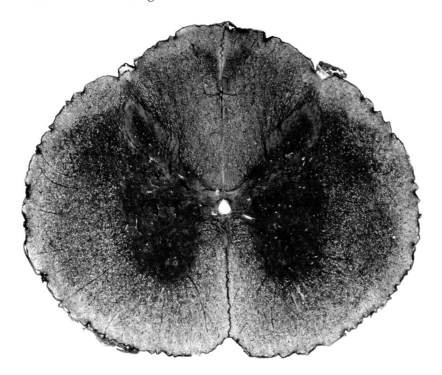

- Cartilage is a special type of connective tissue made of the strong protein collagen. This makes cartilage very flexible and bendy. You have cartilage in your ears and in the tip of your nose. These body parts are softer and more bendable than where your bones are.

- A typical adult has 206 different bones in his body. To show the range, the largest is the femur in the leg, which is just under twenty inches on average, while the smallest bones are found inside our ears at only a few millimeters.

- Newborn babies have somewhere between 270 and 300 bones when they are born. This is because their skeleton is not fully formed. As the baby grows, some of his bones start to join and fuse together to become a single bone, which explains why they would have more bones than an adult. One example of this can be observed in the head of a newborn baby. The human skull is made of several bones that fuse, or join together. In a newborn baby, those bones are separate. The spaces between the bones are filled with cartilage. The most prominent of these spaces is the "soft spot," also called the fontanelle, on top of the baby's head. If you were to gently rub your fingers over a baby's head, you may be able to feel the spot where the bone has not yet grown together. This makes it much easier for the baby to be born because his head can change shape a little bit.

- The human body has over six hundred different muscles.
- Skeletal muscles help us move because they contract, or shorten, which moves our bones, which helps us move.
- The reason a turkey has what we call "white meat" and "dark meat" is because its muscles are used for different things. The turkey breasts are muscles that move the bird's wings. Turkeys do not fly much, so these muscles are used only when they are needed, like when the bird is trying to escape from a hunter or predator. They can contract very quickly, but they also tire quickly because they have low amounts of mitochondria and a smaller blood supply, resulting in a lighter color (less blood = lighter color). The thighs of the turkey, meanwhile, are used all the time because the turkey primarily stands and walks. These muscles must be able to work for long periods of time without becoming tired. As a result, they have high amounts of mitochondria and a rich supply of blood. This is what gives the thighs a darker color.
- The largest cell we know of is a neuron from a Giant Squid. This nerve stretches from the head of the squid all the way down to the end of its tentacles, making it the length of the squid, or about forty feet long!
- The most prominent organ is the brain, which is estimated to contain over eighty billion neurons!

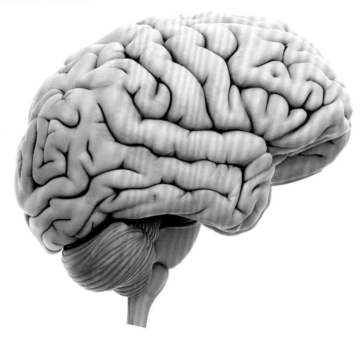

- In an adult, the spinal cord is about eighteen inches long and has a diameter of about one centimeter.
- There are hundreds of millions of nerves in our body. If you put them all together, they would reach around forty-five miles in length!
- What we call the "funny bone" is actually a nerve (the ulnar nerve).
- The sciatic nerve is the longest and thickest nerve in the human body, running from the lower back all the way down the leg to the foot.

- The sense of taste is largely affected by our sense of smell. If you ever had a cold with a stuffy nose, you probably noticed that your food does not taste as good because you can't smell.
- A protein called hemoglobin found inside the red blood cells is what gives blood its red color.
- Each red blood cell contains about 250 million hemoglobin molecules; this means that each cell carries about one billion oxygen molecules!
- *Anyone* can receive Type O blood because it is considered the universal donor, meaning everyone's body can receive it.
- While at rest, a child usually has a heart rate between eighty and ninety beats per minute. But when you play or exercise, your heart rate will increase (become faster). This is because the heart needs to move more blood through the blood vessels to deliver more oxygen to the muscles.
- Maximum heart rate is 220 minus your age. So if you are ten, your maximum heart rate would be 210 and your target heart rate would be 135–158 beats per minute.
- The entire journey of one red blood cell from the heart moving all throughout the body and back to the heart takes place in just forty-five seconds.
- Air enters our bodies through our nose and gets warmed up and moisture is added to it so that it will not damage the tissues that line the respiratory system. The blood in the capillaries is warm and helps to warm up the air. The air becomes moist by the mucus (or snot) that lines the inside of your nasal cavity. When you breathe in through your mouth, the air does not get warmed or humidified. This is why when you have a cold and can't breathe through your nose, you end up having a sore throat. The cold, dry air dries out the back of your throat, causing it to be sore.
- Inside the voice box are folds of throat tissue that form the vocal cords. As air passes over the vocal cords, they vibrate to produce sound. The pitch—that is, the highness or lowness—of the sound depends on how tight the vocal cords are stretched. Vocal cords that are more tightly stretched produce higher sounds while vocal cords that are more relaxed produce lower sounds.

*Remember:*
*It's important to get plenty of exercise and eat a good diet (fruits, vegetables, meats, etc.) if we want to keep our bodies healthy.*

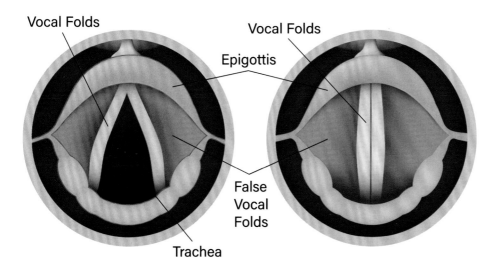

- At the ends of the smallest bronchioles are tiny air sacs called alveoli. Each alveolus is surrounded by a network of capillaries. As the blood moves through the capillaries, oxygen moves out of the alveoli and into the blood. At the same time, the carbon dioxide in the blood moves out of the blood and into the alveoli. While the alveoli themselves are very tiny (you can only see them under the microscope), if you were to take all the alveoli together and lay them out flat, they would cover most of a tennis court. This means that there is a lot of area for the oxygen and carbon dioxide gases to cross into and out of the blood.

- Most people breathe between fifteen and twenty-five times in a minute. Children and women breathe a little bit faster than men. This is called the respiration rate and is controlled by the respiration center in the brain stem.

- Our ability to control how often we breathe is limited. This is because the body is able to sense how much oxygen and carbon dioxide is in the blood. If there is too much carbon dioxide in your blood, special chemical sensors send a message to the brain stem. The brain stem then sends a message to the diaphragm telling it to contract so that you will breathe. This is what happens if you try to hold your breath too long. Your brain tells your body that you need to get rid of the carbon dioxide and forces you to take a breath.

- There are thousands of different enzymes in our bodies, each with a particular job.

- When we put food into our mouths, or if we smell some cookies baking in the oven, it causes our mouth to begin to fill with a watery substance called saliva. The saliva has enzymes that break down the food through chemical reactions. If you put a small soda cracker or piece of white bread and place it in your mouth, and let it sit on your tongue for a few minutes, you will notice that the food starts to get soft and dissolve. This is because the enzymes in your saliva are chemically breaking down the food, even without you chewing it.

- Our stomach juices are more acidic than soda, coffee, or orange juice!

- It is not uncommon for your stomach to "growl" or rumble when you are hungry. The noise is caused by the movement of food, fluid, and mucus through your stomach and intestines. When you're hungry, your brain sends a message to the muscular tissue in the walls of the digestive organs. The muscles start contracting, which causes the noise. Grumbling noises can also keep happening after you eat. As the muscles contract, they continue to push and squeeze the digested food through your digestive tract, producing more noise.

- If you were to stretch out the entire length of an adult's small intestine, it would stretch about twenty feet! So why do we call it small? Because it has a smaller diameter (about an inch) compared to the large intestine (about 2.5 inches).

*Cells & Systems Fun Fact:*
*At only a few days after conception, a human child in the womb is the size of only a tiny poppyseed!*

- For many years, scientists though the appendix had no purpose. Now we know that it provides a protective home for good bacteria that live inside our intestines. Yes, you have bacteria living inside you! These bacteria help to break down some of the food products that our bodies can't digest on their own. The presence of these bacteria also helps keep out bad bacteria that might make you sick.

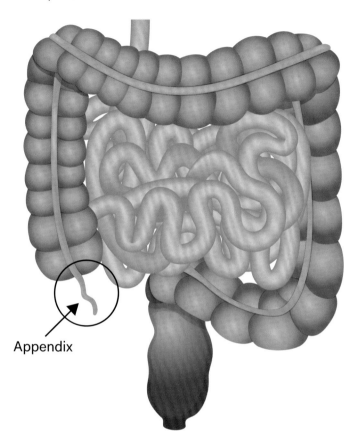

Appendix

- A properly working kidney can filter about a half a cup of blood every minute. This quickly adds up to about forty-five gallons of blood in the course of one day!
- Amazingly, God made us so that we can live with just one half of one kidney. In very special cases, doctors can take a working kidney from one person and put it into a patient with a non-working kidney.
- At only a few days after conception, a human child in the womb is the size of only a tiny poppyseed!
- You will keep on growing until you get to be around twenty years old, and then most people stop getting taller. But they still have a lot of life left to live. A healthy person can live well past eighty years old, and some people live to be one hundred!

# KEY TERMS

**Actin** – *Chapter 6*: Thin protein filaments found in myofibrils that work with thicker filaments (myosin) to help muscles contract, or move.

**Alveoli** (singular alveolus) – *Chapter 9*: Tiny air sac in the lungs where the blood exchanges oxygen and carbon dioxide during the process of breathing in and out.

**Antagonistic pairs** – *Chapter 6*: Muscles that work opposite each other, such as the biceps and the triceps.

**Aorta** – *Chapter 8*: The biggest and most primary artery in the body.

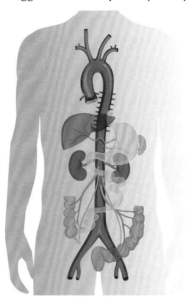

*Cells & Systems Fun Fact:*
*Most people breathe between fifteen and twenty-five times in a minute. Children and women breathe a little bit faster than men. This is called the respiration rate and is controlled by the respiration center in the brain stem.*

**Appendicular skeleton system** – *Chapter 5*: One of the two major parts of the human skeletal system; includes all of the other bones outside of the skull and vertebral column (the axial skeletal system).

**Appendix** – *Chapter 10*: A small fingerlike tail off the end of the large intestine near where it attaches to the small intestine that provides a protective home for good bacteria that live inside the intestines.

**Arteries** – *Chapter 8*: Blood vessels that carry blood *away* from the heart; they have thick walls because they must be able to withstand the pressure of the blood pushing against them as the blood leaves the heart.

**Arterioles** – *Chapter 8*: Small arteries.

**Asthma** – *Chapter 9*: A medical condition of the lungs where the air passageways such as the bronchi and bronchioles constrict or get smaller, causing shortness of breath, coughing, and wheezing.

**Atrium/Atria** – *Chapter 8*: The two chambers of the heart that *receive* the blood from the veins as they return the blood back to the heart.

**Atrophy** – *Chapter 6*: The wasting away or breaking down of tissues, organs, and muscles in the body.

**Auditory nerve** – *Chapter 7*: A nerve that sends signals from the inner ear to the temporal lobe of the brain.

**Auricle** – *Chapter 7*: The visible portion of the external ear directing sound waves into the ear canal; makes up the outer ear along with the tympanic membrane and the ear canal.

**Axial skeletal system** – *Chapter 5*: One of two major parts of the human skeletal system, includes the skull and vertebral column.

**Axon** – *Chapter 7*: A single long fiber that extends off the cell body in a neuron and carries stimuli, or "information," picked up by the dendrites away from the cell body to other cells or to a muscle.

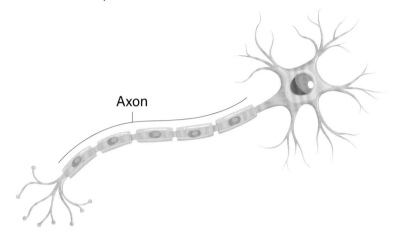

**Bacteria** – *Chapter 2*: Small organisms (living things) made of a single cell, unseen by the naked eye; some bacteria can make you sick (like germs), but some can actually be good for you.

**Bile** – *Chapter 10*: A chemical produced by the liver that helps break down fats in food.

**Bladder** – *Chapter 11*: A balloon-like organ in the lower abdomen that stores urine waste.

**Blood vessels** – *Chapter 8*: Tubes (arteries, veins, capillaries) through which the blood circulates throughout the entire body, carrying oxygen and nutrients to all the different cells.

**Bones** – *Chapter 5*: Living tissues made of thousands of cells called osteocytes; they make up the human skeletal system.

**Brain stem** – *Chapter 7*: The third part of the brain; connects the brain and spinal cord and so passes information between them to send messages to and from the rest of the nervous system.

**Bronchi** (singular bronchus) – *Chapter 9*: Two large windpipes that carry air from your trachea into your lungs (one into each lung).

**Cancer** – *Chapter 3*: A type of disease caused by uncontrolled cell growth.

**Capillaries** – *Chapter 8*: The smallest blood vessel in the human body, found in all our muscles, skin, and tissues.

**Cardiac muscle** – *Chapter 6*: A type of involuntary muscle, this tissue makes up the heart.

**Cardiovascular system (circulatory system)** – *Chapter 8*: Consists of the heart, the blood, and the blood vessels; its function is to transport nutrients and oxygen-rich blood to all parts of the body and carry deoxygenated blood back to the lungs.

**Cartilage** – *Chapter 5*: A special type of connective tissue made of the strong protein collagen, which makes it very flexible and bendy; cartilage structures form the frame for bone to develop and grow.

**Cell** – *Introduction*: The basic building blocks of life; they make up all living organism and the tissues of our bodies.

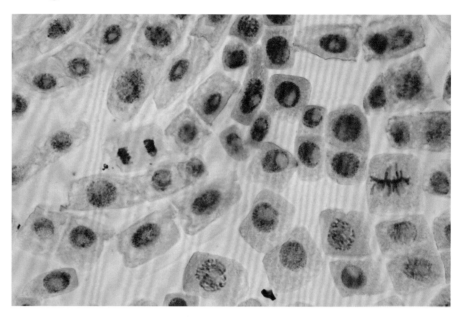

**Cell body** – *Chapter 7*: The central part of a neuron that contains the nucleus and other organelles.

**Cell division** – *Chapter 3*: The process by which a cell divides itself and reproduces (makes more of itself).

**Cell membrane** – *Chapter 1*: An outer skin of sorts that holds the cell together and controls what moves in and out of the cell; sometimes also called the *plasma membrane*.

**Cell Theory** – *Chapter 1*: Theory summarizing the basic facts about cells, formed over many years by various scientists; states that (1) all living things are made of cells; (2) cells are the basic unit of structure and function for all living things; and (3) all cells come from pre-existing cells.

**Cell wall** – *Chapter 2*: Found outside of the cell membrane in plant cells and bacterial cells; made of proteins and other molecules that help to protect the cell from the outside environment.

**Cellular respiration** – *Chapter 2*: The process by which special enzymes inside the mitochondria break down sugar molecules into energy for the cell. This process requires oxygen.

**Cellulose** – *Chapter 2*: A complex sugar molecule that makes up the cell walls of plants, giving it support and protection.

**Central nervous system** – *Chapter 7*: One of the two parts of the human nervous system; includes the brain and spinal cord.

**Cerebellum** – *Chapter 7*: The second part of the brain found beneath the occipital lobe; its main role is to help with balance and coordinating our movements.

**Cerebrum** – *Chapter 7*: The largest part of the brain; divided into two halves, or hemispheres, that each control various functions such as speech, thought, emotions, memory, touch, and more.

**Chemoreceptors** – *Chapter 7*: A sensory cell or organ responsive to chemical stimuli; they help with our sense of taste and smell.

**Chemotherapy** – *Chapter 3*: A type of medication for cancer patients that kills the bad cancer cells.

**Chloroplasts** – *Chapter 2*: Tiny structures—organelles—in plant cells that capture light energy from the sun to carry out the process of photosynthesis.

**Chromosome** – *Chapter 2*: A thread-like structure found inside the nucleus of a cell that carries genetic information from one generation to another.

**Chyme** – *Chapter 10*: Acidic fluid consisting of digested food and stomach juices that passes from the stomach to the small intestine.

**Cochlea** – *Chapter 7*: A snail-shaped structure in the inner ear that is filled with fluid; when vibrations enter the ear, it causes waves in the fluid, which bend small sensory cells, causing them to send a signal through nerves in the inner ear to the temporal lobe of the brain.

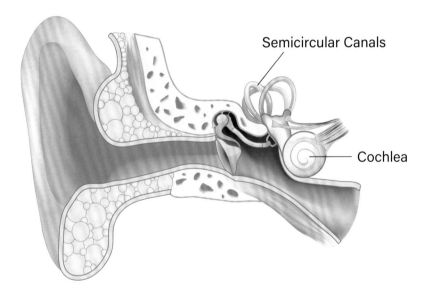

**Collagen** – *Chapter 5*: A structural protein found in skin and other connective tissues; collagen fibers are elastic and bendy and can stretch, giving our skin and bones this same quality.

**Compound microscope** – *Chapter 1*: An instrument that uses more than one lens—known as *converging lens*—to make objects look bigger than they really are, allowing for scientific observation and study.

**Connective tissue** – *Chapter 4*: One of the four types of tissues found in our bodies; these tissues connect, bind, or support.

**Converging lens** – *Chapter 1*: The type of lens used in a *compound microscope*.

**Cornea** – *Chapter 7*: The clear surface covering the outside of the eye where light first passes through.

**Cytoplasm** – *Chapter 1*: A gel-like fluid made mostly of water that fills the inside of the cell; contains the cells organelles.

**Daughter cells** – *Chapter 3*: The resulting cells from a single cell dividing itself.

**Dendrite** – *Chapter 7*: Finger-like extensions coming off the cell body in a neuron that pick up stimuli, or "information," from other cells or from the environment.

**Dermis** – *Chapter 4*: The bottom layer of our skin, much thicker than the epidermis; it contains special cells called nerves as well as our blood vessels, sweat glands, and hair follicles.

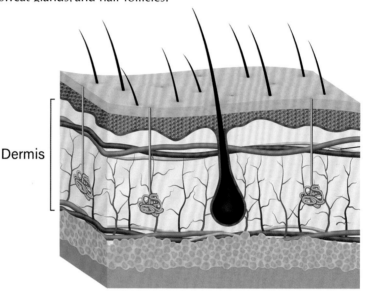

**Diabetes** – *Chapter 10*: A disease when the body cannot keep the blood glucose levels steady.

**Diaphragm** – *Chapter 9*: A thin muscle the lungs sit on top of, separates the chest cavity from the stomach cavity; helps regulate breathing.

**Digestive system** – Chapter 10: Its primary function is to break down food into smaller components until they can be absorbed, assimilated into the body, and passed out as waste.

**DNA (Deoxyribonucleic acid)** – *Chapter 1*: The genetic material inside each cell that contains the "instructions" that tell the cell what organelles to build and how to carry out its job (like the directions that make you who you are).

**DNA replication** – *Chapter 3*: The process by which, during cell division and reproduction, the DNA within the cell makes a complete copy of itself.

**Down syndrome** – *Chapter 3*: A condition where an extra chromosome is found in a human cell (forty-seven instead of forty-six) caused by a mistake in the cell division process.

**Ear canal** – *Chapter 7*: The small tube that opens into the ear and helps funnel sound waves inside; makes up the outer ear along with the auricle and the tympanic membrane.

**Embryonic stage** – *Chapter 12*: The first two months of human development inside the mother's womb, beginning with the rapid division of cells and through the early development of major organs.

**Enzymes** – *Chapters 2 & 10*: A type of protein that helps carry out chemical reactions in our cell such as breaking down larger molecules into smaller molecules; there are thousands of different enzymes in our cells, each with its own specific shape and function.

**Epidermis** – *Chapter 4*: The outer layer of our skin made up of many layers of cells.

**Epiglottis** – *Chapter 9*: An internal spoon-shaped piece of cartilage that covers the opening to the trachea; prevents food or drink from entering into the respiratory system.

**Epithelial tissues** – *Chapter 4*: One of the four types of tissues found in our bodies; these tissues form the cover for our bodies and organs, like our skin.

**Erythropoietin** – *Chapter 11*: A special hormone made in the kidneys that carries a message to the bones telling them to make more red blood cells.

**Esophagus** – *Chapter 10*: A muscular tube that connects the throat with the stomach.

*Cells & Systems Fun Fact:*
*The entire journey of one red blood cell from the heart moving all throughout the body and back to the heart takes place in just forty-five seconds.*

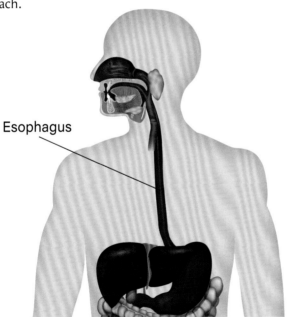

**Eukaryotic cells** – *Chapter 2*: A type of cell that has its DNA enclosed in a membrane bound nucleus; they tend to be larger and more complex than prokaryotic cells.

**Fascicles** – *Chapter 6*: A bundle of structures, such as nerve or muscle fibers.

**Fetal stage** – *Chapter 12*: The last seven months of human development inside the mother's womb.

**Flagellum** – *Chapter 2*: A tail-like structure in some prokaryote cells that rotates or spins, helping the cell to move.

**Flat bones** – *Chapter 5*: Bones shaped more like a plate; examples are found in the skull and ribs.

**Fontanelle** – *Chapter 5*: The "soft spot" on newborn babies' skulls; the result of bones in the baby not being forged together yet, but instead being filled in between with cartilage.

**Gallbladder** – *Chapter 10*: A sac beneath the liver that stores bile.

**Gene** – *Chapter 12*: A unit or segment of DNA that plays a vital role in *inheritance*.

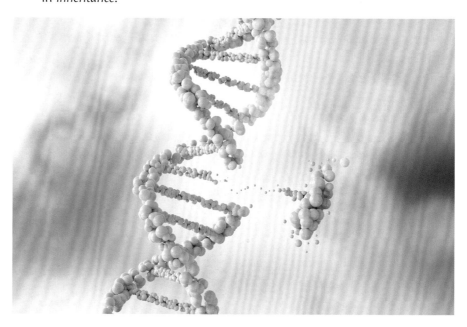

**Gene expression** – *Chapter 12*: The process by which the DNA in a cell is read so the cell knows how to perform its function.

**Glands** – *Chapter 4*: Tiny sacs in the bottom layer in our skin (the dermis) that help us cool off by producing sweat.

**Glomerulus** – *Chapter 11*: A ball-shaped group of capillaries inside the kidneys where waste products are filtered from the blood.

**Growth plate** – *Chapter 5*: A thin layer of cartilage, usually near the end of long bones, where much of the bone growth takes place from new cells forming.

**Hair follicles** – *Chapter 4*: Tiny structures in the bottom layer of our skin (the dermis) from which hair grows.

**Heart rate** – *Chapter 8*: How many times the heart beats per minute; can vary based on factors like age and whether or not the body is at rest or moving.

**Hemoglobin** – *Chapter 8*: A protein found in red blood cells that carries four molecules of heme that hold an iron atom; it is the heme that binds to an oxygen molecule and gives red blood cells their red color.

**Hormone** – *Chapter 11*: A special protein that carries a message from one organ to another organ.

**Hydrophilic** – *Chapter 2*: Water-loving.

**Hydrophobic** – *Chapter 2*: Water-fearing.

**Hypodermis** – *Chapter 4*: A layer in our skin below the dermis where fat tissue is found, helping to keep us warm.

**Inheritance** – *Chapter 12*: The passing of traits from parents to children.
**Ingestion** – *Chapter 10*: The taking in of food through the mouth.
**Intercostal muscles** – *Chapter 9*: Groups of muscles that run between the ribs forming the chest wall; they help control breathing.
**Involuntary muscles** – *Chapter 6*: Muscles which move on their own without us intentionally thinking about them; these include cardiac muscle and smooth muscle.
**Iris** – *Chapter 7*: The colored tissue part of the eye made of muscles that control the size of the pupil.

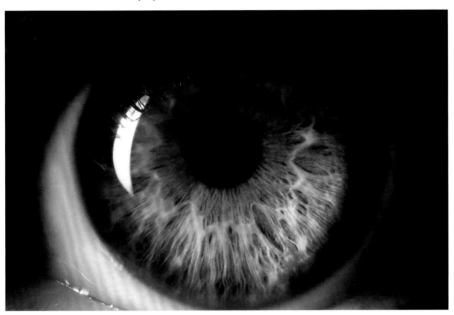

**Irregular bones** – *Chapter 5*: Bones which have odd shapes, like the vertebrae that make up the backbone.
**Joint** – *Chapter 5*: The area where two or more bones comes together.
**Keratin** – *Chapter 4*: A special protein that protects our skin and keeps it from tearing.
**Kidneys** – *Chapter 11*: Bean-shaped organs found along either side of your backbone in your lower back that filter waste products out of the blood.
**Lacunae** – *Chapter 5*: Little "windows" inside the osteon that house the osteocytes (bone cells).
**Large intestine** – *Chapter 10*: A tube-like organ connected to the small intestine and the anus that helps the digestive system move food through the body.
**Larynx** – *Chapter 9*: Also known as the voice box, a tubular structure connecting the pharynx to the *trachea* (windpipe).
**Lens** – *Chapter 7*: A gelatin-filled sphere held in place by small muscles inside your eye; these muscles can stretch the lens flat when you are looking at something far away, but when you look at something close, the lens gets fat and round; the lens changes shape to help focus the light on the back of the eye.

**Ligaments** – *Chapter 5*: Long bands of connective tissue that hold joints together and keep bones from slipping out of place.

**Liver** – *Chapter 10*: The large organ found on the right side of the abdomen just below the ribs; helps remove toxins from the body and through the production of *bile* helps break down fats in food.

**Long bones** – *Chapter 5*: Longer bones that are hollow cylinders with knobs on their ends, inside which rests a substance called yellow marrow, which stores fat; they also contain spongy bone, where blood is made; examples include bones found in the legs, arms, fingers, and toes.

*Remember:*
*Our stomach juices are more acidic than soda, coffee, or orange juice!*

**Magnifying glass** – *Chapter 1*: An instrument that makes use of a curved piece of glass—a lens—to enlarge an object for better observation; the curved glass bends light as it passes through, changing how we see the object on the other side of the lens (it becomes bigger).

**Meiosis** – *Chapter 3*: A unique kind of cell division in certain kinds of plant and animal cells where the cells will divide two times in a row to make four cells with half the number of chromosomes.

**Melanin** – *Chapter 4*: A special pigment found in our skin (in the epidermis) that gives color to our hair and skin.

**Mitochondria** – *Chapter 2*: Small bean-shaped structures (organelles) in eukaryotic cells that break down sugars to make energy for the cell; sometimes called the "powerhouse" of the cell because of how it makes energy.

**Mitosis** – *Chapter 3*: The process by which eukaryotic cells reproduce.

**Motor neurons** – *Chapter 7*: Neurons that carry information to a muscle or other organ that then responds to the stimulus.

**Muscular system** – *Chapter 6*: Made up of the muscles in our bodies which support us and help us move, as well as stabilizing joints and allowing us to keep a steady body temperature.

**Muscular tissue** – *Chapter 4*: One of the four types of tissues found on our bodies; these tissues make up our muscles.

**Myelin sheath** – *Chapter 7*: A protective layer of cells that wraps around the axon of a neuron to provide insulation (protection), allowing nerve signals to travel more quickly.

**Myofibrils** – *Chapter 6*: Rod-like muscle protein fibers, or organelles, composed of long proteins that work together to help the muscle contract, or move.

**Myosin** – *Chapter 6*: Long, thick, golf club-shaped protein filaments found in myofibrils that work with thin protein filaments (actin) to contract, or move, the muscle.

**Nasal cavity** – *Chapter 9*: The air-filled space and skin within the nose.

**Nephrons** – *Chapter 11*: Tiny tubes inside the kidneys that act as filtering units.

**Nerves** – *Chapters 4 & 7*: Special cells, bundles of neurons, found in the bottom layer of our skin (the dermis) that help us feel; they run throughout the human body forming a network, allowing the different parts of the body to communicate with each other.

**Nervous system** – *Chapter 7*: Allows us to *sense* things in our world, register them, and then respond in an appropriate way; composed of the central and peripheral nervous systems.

**Nervous tissue** – *Chapter 4*: One of the four types of tissues found on our bodies; these tissues work together to form our nervous system.

**Neurons** – *Chapter 7*: The main cell that makes up all the structures in the nervous system.

**Nondisjunction** – *Chapter 3*: A kind of mistake that sometimes happens during cell division where chromosomes do not separate properly.

**Nucleus** – *Chapter 2*: Rests at the center of the cell, a membrane-bound organelle that contains the cell's chromosomes.

**Optic nerve** – *Chapter 7*: A special nerve behind the eye which carries the message to the occipital lobe of the brain.

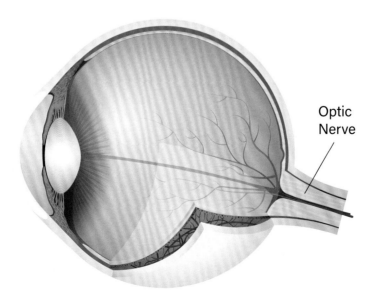

**Osteocytes** – *Chapter 5*: Bone cells.

**Osteon** – *Chapter 5*: Circular structures of minerals found in our bones that house the osteocytes (bone cells), blood vessels, and nerves.

**Pancreas** – *Chapter 10*: An organ found on the left side of the abdominal region that releases several different enzymes to help to break down sugars and proteins; also functions to help the body keep the right amount of sugar (glucose) in the blood.

**Peripheral nervous system** – *Chapter 7*: One of two parts of the human nervous system; includes all the areas outside the brain and spinal cord.

**Pharynx** – *Chapter 9*: The back of the throat which is shared by the respiratory and digestive systems.

**Phospholipids** – *Chapter 2*: A type of molecule possessing a "head" and two "tails" that makes up the cell membrane; it is unique because it is both hydrophilic (the head is water-loving) and hydrophobic (the tails are "water-fearing"), an attribute that performs special functions for the cell, like determining what passes in and out of the cell.

**Photoreceptors** – *Chapter 7*: Special sensory cells in the eye that respond to light.

**Placenta** – *Chapter 12*: A "temporary" organ that develops in the mother's uterus during pregnancy to provide oxygen and nutrients to the growing baby.

**Plasma** – *Chapter 8*: The watery part of the blood.

**Platelets** – *Chapter 8*: Tiny fragments of cells in the plasma that clump together to stop bleeding.

**Prokaryotic cells** – *Chapter 2*: A type of cell that has its DNA floating freely in the cytoplasm; they tend to be smaller and simpler than eukaryote cells. A bacteria cell is an example of a prokaryotic cell.

**Proteins** – *Chapter 2*: Large, complex molecules that play many critical and beneficial roles for our bodies, including allowing substances to pass in and out of cells through the cell membrane, breaking down foods, and serving as the building blocks for making other structures in cells and in our bodies, such as muscle and hair. They are made of amino acids.

**Pupil** – *Chapter 7*: The small black circle at the center of the eye; an open window that allows light to enter into the eye.

**Red blood cell** – *Chapter 8*: Type of cell in the plasma that carries oxygen; a protein called hemoglobin is inside them which gives blood its red color.

**Reflex** – *Chapter 7*: An involuntary response to a stimulus.

**Renal artery** – *Chapter 11*: The main blood vessel that supplies blood to a kidney.

**Reproduce** – *Chapter 3*: When a living organism makes more of itself.

**Respiration rate** – *Chapter 9*: The rate at which a person breathes per minute.

**Respiratory system** – *Chapter 9*: The system in the human body that takes oxygen gas into your lungs so that it is available for the blood to pick up and carry to the rest of your body; includes your nose, mouth, throat, lungs, and other structures associated with them.

---

*Remember:*

*Keeping our kidneys healthy by drinking plenty of water is important because the kidneys filter waste and toxins out of our blood.*

**Retina** – *Chapter 7*: The inner layer of the eye which contains the special sensory cells called *photoreceptors* that respond to light.

**Ribosomes** – *Chapter 2*: Small structures inside the cell that make proteins.

**Saliva** – *Chapter 10*: A watery substance in the mouth which contains enzymes that break down food through chemical reactions.

**Sarcomere** – *Chapter 6*: A unit of striated muscle tissue composed of thick and thin protein filaments (myosin and actin).

**Sensory neurons** – *Chapter 7*: Neurons that sense information from the environment.

**Short bones** – *Chapter 5*: Shorter bones that are approximately as wide as they are long; examples are found in the wrists and ankles.

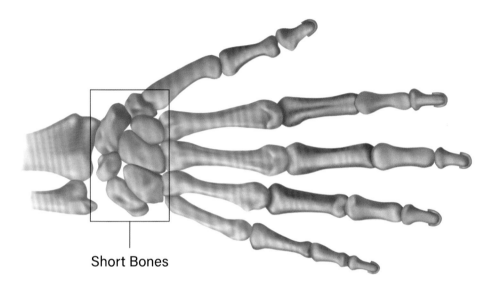

Short Bones

**Skeletal muscles** – *Chapter 6*: Muscles that work with our bones to help us move.

**Skeletal system** – *Chapter 5*: The collection of bones that make up the framework of the human body.

**Small intestine** – *Chapter 10*: The place in the digestive system where all the nutrients that have been broken down get absorbed into your body. It connects the stomach to the large intestine.

**Smooth muscle** – *Chapter 6*: An example of an involuntary muscle, these make up the walls of many of our digestive organs, such as our stomach and intestines.

**Spinal cord** – *Chapter 7*: Along with the brain, a part of the central nervous system; found inside the vertebrae that make up the backbone; nerves extend off the spinal cord and carry information between the brain and other parts of the body.

**Spongy bone** – *Chapter 5*: A type of lighter and less dense bone consisting of networks of bone tissue with a lot of small spaces in between, where marrow and blood vessels are found.

**Stomach** – *Chapter 10*: Like a storage bag for your food; holds, stores, and breaks down the food we eat.

**Tissues** – *Chapter 4*: A collection of similar cells working together for the same purpose.

**Trachea** – *Chapter 9*: The windpipe.

**Tympanic membrane** – *Chapter 7*: Also called the eardrum, a membrane that makes up part of the outer ear along with the auricle and ear canal; it vibrates in response to sound waves.

**Ureter** – *Chapter 11*: A small tube connecting the kidneys to the bladder.

**Urethra** – *Chapter 11*: A small tube that releases urine out of the bladder.

**Urinary system** – *Chapter 11*: Consists of the kidneys, bladder, urethra, and ureter; its primary function is to eliminate waste through urine and filter the blood.

**Valve** – *Chapter 8*: Flap-like structures (like doors) that separate atria and the ventricles; they control the movement of blood through the heart.

**Vena cava** – *Chapter 8*: The largest vein in the human body which dumps blood back into the heart.

**Veins** – *Chapter 8*: Blood vessels that *return* blood to the heart.

**Ventilation** – *Chapter 9*: The movement of air in and out of your lungs; it consists of two steps, inhalation, or bringing air into the lungs, and exhalation, moving air out of the lungs.

**Ventricles** – *Chapter 8*: The two chambers of the heart that *force the blood out* of it when they contract.

**Venules** – *Chapter 8*: Small veins.

**Villi (singular *villus*)** – *Chapter 10*: Fingerlike structures lining the small intestine that increase the surface area of the small intestines and result in the absorption of more nutrients.

**Vocal cords** – *Chapter 9*: Folds of throat tissue that vibrate to produce sound when air passes over them.

**Voluntary muscles** – *Chapter 6*: Muscles that we can move by intentionally thinking about them, as opposed to *involuntary muscles*, which move on their own without us thinking about them.

**White blood cell** – *Chapter 8*: A type of cell suspended in the plasma that helps fight off infections.

---

*Cells & Systems Fun Fact:*

*The human body has over six hundred different muscles!*

# IMAGE CREDITS

Front cover  DNA, 3D illustration, akr11_ss / white blood cells, Shilova Ekaterina / Stem Cells Immunotherapy, CI Photos / Red Blood Cells, Rashevskyi Viacheslav / Lomitapide cholesterol lowering drug molecule, StudioMolekuul / Reaching hands from The Creation of Adam of Michelangelo, Freeda Michaux © Shutterstock.com

pI  Anatomy 3D Illustrations. © adike, Shutterstock.com
pII  Micrograph of blood vessel, artery and vein. © Pan Xunbin, Shutterstock.com
pV  Space background with many stars © Sundays Photography, Shutterstock.com
pVI  Fluorescence microscopic view of human skin cells in culture. © Vshivkova, Shutterstock.com
pVII  Title. © Artist, Shutterstock.com
pVIII  Medically accurate illustration of the femur. © SciePro, Shutterstock.com
pIX  Streaming blood cells isolated on white. © martan, Shutterstock.com
pX  3D rendered medically accurate illustration of the human cell anatomy. © SciePro, Shutterstock.com
pXII  Engraving from Micrographia, 1665, by Robert Hooke. Source: https://wellcomecollection.org/works/y5tm56u5. License: (CC BY 4.0), https://creativecommons.org/licenses/by/4.0/
p1  Microscopes, 17th Century, Hooke's compound microscope and its illuminating system, 1665-1675 / Science Museum, London, UK / Photo credit: SSPL/UIG / Bridgeman Images
p2  Robert Hooke's microscope. From Scheme I. of his 1665 Micrographia. Date: 1665. Source: w:en:Image:Hooke-microscope.png Author: Robert Hooke [public domain], via Wikimedia Commons
p2  Portrait of Sacharias Jansen. Date: 1655. Source: Pierre Borel, De vero telescopii inventore Author: Pierre Borel [public domain], via Wikimedia Commons
p2  Converging lens diagram. © Drawing For Freedom, Shutterstock.com
p3  Engraving of an ant in Micrographia, 1665, by Robert Hooke. Source: https://wellcomecollection.org/works/te9yj73g. License: (CC BY 4.0), https://creativecommons.org/licenses/by/4.0/
p3  Robert Hooke's illustration of a flea from Micrographia. Date: 1665. Source: National Library of Wales Author: Robert Hooke [public domain], via Wikimedia Commons
p4  Bacterial cell anatomy. © OSweetNature, Shutterstock.com
p4  Egg of ostrich and chicken in egg carton. © Valery Bareta, Shutterstock.com
p5  Professional biological laboratory microscope. © Gerg Ross, Shutterstock.com
p6  Redi's Experiment diagram. © BlueRingMedia, Shutterstock.com
p6-7  Hessian sackcloth woven texture pattern background. © Chinnapong, Shutterstock.com
p7  Statue of Francesco Redi. Source: RicciSpeziari. Photographer: Riccardo Speziari. License: (CC BY-SA 3.0), https://creativecommons.org/licenses/by-sa/3.0/deed.en
p8-9  Various cells. © Kateryna Kon, wacomka, StudioSmart, Athitat Shinagowin, cenksns, Shutterstock.com
p11  Phospholipid / illustrated part of cell membrane. © magnetix, Shutterstock.com
p11  Bacteria colonies from freshwater pond. © Ekky Ilham, Shutterstock.com
p12  Enzyme function diagram. © Designua, Shutterstock.com
p12  Chromosomes. © Designua, Shutterstock.com
p13  Mitochondria, 3D illustration. © Kateryna Kon, Shutterstock.com
p13  Punctated star moss (Rhizomnium punctatum), lamina cells, magnified 400x. Source: Kristian Peters. License: (CC BY-SA 3.0), https://creativecommons.org/licenses/by-sa/3.0/deed.en
p14  Cell cross section structure detailed. © Tefi, Shutterstock.com
p15  Illustration of the bacteria cell structure. © Aldona Griskeviciene, Shutterstock.com
p16-17  The Sermon on the Mount, from the Sistine Chapel, c.1481-83 (fresco) / Rosselli, Cosimo (1439-1507) / Italian / Vatican Museums and Galleries, Vatican City / Bridgeman Images
p18-19  Process division of cell. © Lukiyanova Natalia frenta, Shutterstock.com
p20  Scraped kid's knee. © TY Lim, Shutterstock.com
p20  Mitosis. © Designua, Shutterstock.com
p21  Sea urchin embryo cell division. © yosuyosun, Shutterstock.com
p21  Diagram of pregnant woman with baby. © BlueRingMedia, Shutterstock.com
p22  Meiosis. © Aldona Griskeviciene, Shutterstock.com
p22  Tomatoes. © Tim UR, Shutterstock.com
p23  Neutrophil, mature (developed) white blood cells. © Tewan Banditrakkanka, Shutterstock.com
p23  Microscopic view of acute myeloid leukemia (AML). © Saiful52, Shutterstock.com
p24  St. Peregrine. Source: https://www.anticoantico.com. Artist: Giacomo Zampa, 17th century [public domain], via Wikimedia Commons
p24  Meiotic nondisjunction. The failure of one or more pairs of homologous chromosomes to separate normally during nuclear division. © Aldona Griskeviciene, Shutterstock.com
p25  Division of a cell, mitosis, 3D illustration. © Kateryna Kon, Shutterstock.com
p26-27  Karyotype of Down syndrome (DS or DNS), also known as trisomy 21, is a genetic disorder caused by the presence of all or part of a third copy of chromosome 21. © kanyanat wongsa, Shutterstock.com
p27  Jerome Lejeune was a French Catholic pro-life paediatrician and geneticist, best known for his discovery of the link of diseases to chromosome abnormalities. He developed the karyotype. Source: Fondation Jérôme Lejeune Author: Fondation Jérôme Lejeune License: (CC BY-SA 3.0), https://creativecommons.org/licenses/by-sa/3.0/deed.en
p28-29  Anatomical structure of the skin 3D illustration. © picmedical, Shutterstock.com
p30  Four tissue types. © Designua, Shutterstock.com
p31  Anatomy of the skin and the layers and elements that compose it.. © ilusmedical, Shutterstock.com
p31  Finger with hangnail. © zulfachri zulkifli, Shutterstock.com
p31  Close up male human skin and hair. © Mas Hendra P, Shutterstock.com

p32  Various skin pigments. © Red Confidential, Peter Kotoff, komkrit Preechachanwate, jugulator, charnsitr, PixieMe, Nathalie Speliers Ufermann, Shutterstock.com
p33  A small scratch on the skin of a little boy's leg. © Egoreichenkov Evgenii, Shutterstock.com
p33  Sunburn on skin. © Nick N A, Shutterstock.com
p33  Paint on hands. © pics five, Shutterstock.com
p34-35  Father Damien, Belgian Catholic missionary, on his mission to the leper colony on Molokai, Hawaii, 1873-1889 (colour litho) / Mills, Alfred Wallis (1855-1942) / British / © Look and Learn / Bridgeman Images
p36-37  Medically accurate 3d illustration of a human athlete © SciePro, Shutterstock.com
p37  Woman running - visible anatomy of the skeleton © SciePro, Shutterstock.com
p38  Human Anatomy full body male skeleton, 3D illustration. © Matis75, Shutterstock.com
p39  Histology of human compact bone tissue under microscope. © Choksawatdikorn, Shutterstock.com
p40  Growing bones illustration: fetus, childhood and adolescence organ development stages scheme. Ossification center and spongy formation © VectorMine, Shutterstock.com
p41  Bone spongy structure close-up. © eranicle, Shutterstock.com
p41  Glass of milk. © Theeradech Sanin, Shutterstock.com
p41  Woman doing pushups showing skeletal structure, 3d illustration. © SciePro, Shutterstock.com
p42  Classification of Bones By Shape. Source: BruceBlaus Author: BruceBlaus License: (CC BY 3.0), https://creativecommons.org/licenses/by/3.0/deed.en
p44  Human Skeleton System Appendicular and Axial Skeleton Anatomy. 3D illustration. © Magic mine, Shutterstock.com
p45  The structure of the human knee joint. Lateral view. © studiovin, Shutterstock.com
p45  Skeleton. © Alesandro14, Shutterstock.com
p46  The Vision of Ezekiel: the Valley of Dry Bones / Stanhope, John Roddam Spencer (1829-1908) / English / Photo © Christie's Images / Bridgeman Images
p48  Anatomical illustration of a runner show muscle structure. © Mopic, Shutterstock.com
p49  Medically accurate 3d illustration of a football player muscle structure © SciePro, Shutterstock.com
p50  Biceps, Triceps - movement of the arm and hand muscles © stihii, Shutterstock.com
p50  Anatomy of the heart with the main arteries and veins that compose it, highlighting the left coronary artery. © ilusmedical, Shutterstock.com
p50  3 types of muscle tissue and cell. © Aldona Griskeviciene, Shutterstock.com
p51  Skeletal striated muscle tissue under the microscope. Muscle fibers. © BioFoto, Shutterstock.com
p51  Chicken breast meat. © Spayder pauk_79, Shutterstock.com
p51  Raw chicken legs. © Sergey Eremin, Shutterstock.com
p52  A sarcomere is the complicated unit of striated muscle tissue. It is the repeating unit between two Z lines, Skeletal muscles are composed of tubular muscle cells(myocytes called muscle fibers or myof. © Akor86, Shutterstock.com
p53  Rowing team. © Dmitrydesign, Shutterstock.com
p55  Male and Female muscular system. © SciePro, Shutterstock.com
p56  Young brown hen. © Olhastock, Shutterstock.com
p57  Blooming potato field with sun and trees. © Elenamiv, Shutterstock.com
p57  St Catherine of Siena, 1888, by Alessandro Franchi (1838-1914). detail. / Franchi, Alessandro (1838-1914) / Italian / © A. Dagli Orti / © NPL - DeA Picture Library / Bridgeman Images
p58  Medically accurate illustration of basketball player's nervous system © SciePro, Shutterstock.com
p59  Woman athlete nervous system, 3D illustration. © S K Chavan, Shutterstock.com
p60  Panorama Tranquility Bay © Vibrant Image Studio, Shutterstock.com
p61  Neuron cell model, 3D rendered illustration. © MattLphotography, Shutterstock.com
p61  Squid © Jiang Zhongyan, Shutterstock.com
p62  Human brain anatomy structure / illustration © miha de, Shutterstock.com
p63  The vertebral column. © Alila Medical Media, Shutterstock.com
p64  Ulnar nerve medical illustration © Chu KyungMin, Shutterstock.com
p65  Human eye anatomy illustration. © Natee Jitthammachai, Shutterstock.com
p66  Ear anatomy illustration. © Medical Art Inc, Shutterstock.com
p67  Girl's nose and tongue. © pathdoc, Shutterstock.com
p68  Christ Healing the Blindman. Artist: Gerardus Duyckinck I. Source: https://www.metmuseum.org/art/collection/search/641448. Credit: Friends of the American Wing Fund, 2014. License: CC0 1.0 Universal (CC01.0), https://creativecommons.org/publicdomain/zero/1.0/deed.en
p69  Saint Lucy of Syracuse, Italy Source: Serse82 Author: Serse82 [public domain], via Wikimedia Commons
p70-71  Human vascular system, 3D rendered illustration. © SciePro, Shutterstock.com
p72  Sacred Heart of Jesus. © Xolopiks, Shutterstock.com
p73  Human Blood. © SciePro, Shutterstock.com
p74  Close up of couperose, red and dilated capillaries, spider veins, vascular problems. © Evgeniya Sheydt, Shutterstock.com
p74  Blood vessel with flowing blood cells, 3D illustration. Small blood vessels, capillaries. © Kateryna Kon, Shutterstock.com
p75  Human arterial and venous circulatory system.. © Olga Bolbot, Shutterstock.com
p76  Medical illustration of human heart. © Chu KyungMin, Shutterstock.com
p76  Reading of one's pulse. © grib_nick, Shutterstock.com
p77  Diagram showing blood flow of the human heart. © GraphicsRF.com, Shutterstock.com
p78  Red blood cells, Aldona Griskeviciene, white blood cells and platelets, supergalactic © Shutterstock.com
p79  The Last Supper (oil on panel) / Macip, Vicente Juan (Juan de Juanes) (c.1510-79) / Spanish / Bridgeman Images
p79  Eucharistic Miracle of Lanciano. Source: Junior Author: Junior [public domain], via Wikimedia Commons
p80-81  Human Respiratory System Lungs Anatomy. 3D rendered illustration. © Magic mine, Shutterstock.com
p83  Respiratory system diagram. © stockshoppe, Shutterstock.com

# IMAGE CREDITS

p84 Tree without leaves. © Potapov Alexander, Shutterstock.com
p84 Schematic illustration of the apparatus, human respiratory, the lungs are in a warm and flat color, and the left part uncovered, you can appreciate in its entirety the branching of the bronchi. © ilusmedical, Shutterstock.com
p85 X-ray of chest of healthy patient. © oksana2010, Shutterstock.com
p85 Asthma-inflamed bronchial tube © BlueRingMedia, Shutterstock.com
p85 Asthma inhaler. © saltodemata, Shutterstock.com
p86 Diaphragm functions in breathing illustration. © BlueRingMedia, Shutterstock.com
p87 Hand weights. © Happy Stock Photo, Shutterstock.com
p87 Jump rope. © Hurst Photo, Shutterstock.com
p88 Pentecost. Artist: Juan Bautista Maíno. Source: http://www.museodelprado.es/uploads/tx_gbobras/P03286.jpg [public domain], via Wikimedia Commons
p90-91 Human Digestive System Anatomy. 3D rendered illustration. © Magic mine, Shutterstock.com
p92 Partially eaten apple. © New Africa, Shutterstock.com
p93 The location of the gastrointestinal tract in the body, the human digestive system. © Marochkina Anastasiia, Shutterstock.com
p94 Sliced bread with a bite out of it. © Yellow Cat, Shutterstock.com
p94 Stomach illustration. © Macrovector, Shutterstock.com
p95 Digestive system, small intestine. © Marochkina Anastasiia, Shutterstock.com
p96 Human liver anatomy. © BlueRingMedia, Shutterstock.com
p96 Digestive system, large intestine. © Marochkina Anastasiia, Shutterstock.com
p97 Pizza box. © pics five, Shutterstock.com
p98 Teresa of Ávila by Peter Paul Rubens, 1615. Source: David Monniaux. Author: David Monniaux. License: (CC BY-SA 3.0), https://creativecommons.org/licenses/by-sa/3.0/deed.en
p98-99 Eucharist. © Sebastian Duda, Shutterstock.com
p100-101 Urinary system anatomy, 3D rendered illustration. © Magic mine, Shutterstock.com
p102 Kidney anatomy. © ilusmedical, Shutterstock.com
p103 Glomerulus. © sciencepics, Shutterstock.com
p104 Balloon inflation. © exopixel, Shutterstock.com
p104 Urinary system anatomy. © La Gorda, Shutterstock.com
p105 Kidney stone (renal stone , renal calculi) (film x-ray KUB (Kidney - Ureter - Bladder) show left renal stone). © Puwadol Jaturawutthichai, Shutterstock.com
p105 Glass of water. © mylisa, Shutterstock.com
p106 Dirty and clean droplets, infection illustration. © Lina Truman, Shutterstock.com
p107 Wooden confessional. © FCG, Shutterstock.com
p107 Return of the Prodigal Son, 1773 (oil on canvas) / Batoni, Pompeo Girolamo (1708-87) / Italian / Bridgeman Images
p108-109 DNA © MiniStocker, Shutterstock.com
p110 DNA Replication, Protein synthesis, Transcription and translation. Biological functions of DNA. Genes and genomes. Genetic code © Designua, Shutterstock.com
p112 Freckles / chaoss, no freckles / IhorL, widow's peak / Phatthanit, straight hairline / New Africa, free ear lobes / aleks333, attached ear lobes / BLACKDAY, dimples / Wischy, no dimples / Marina Demeshko, artist, straight little finger / Pradyumna prasad upadhyay, tongue roller / Mateusz Kopyt, non-tongue roller / Djomas, right and left handed / TETSUZO KIZZGAWA © Shutterstock.com — Bent little finger. Source: TheFrenchTickler1031. Author: TheFrenchTickler1031. License: (CC BY-SA 4.0), https://creativecommons.org/licenses/by-sa/4.0/deed.en
p113 3D ultrasound of baby in mother's womb. © Valentina Razumova, Shutterstock.com
p114 Human development in the womb. © Pasha Smith, Shutterstock.com
p115 Molecular structure of L - proline (important amino acid), 3D rendering. © Raimundo79, Shutterstock.com
p116 Scanning electron microscopy of human cell surface. Extreme close-up of mammalian cell surface morphology. Erythrocytes, connective tissue and collagen fibres. © thomaslabriekl, Shutterstock.com
p118 Cell division. © Lukiyanova Natalia frenta, Shutterstock.com
p119 Cross section of spinal cord viewed under the microscope. © Choksawatdikorn, Shutterstock.com
p120 3D rendered medically accurate illustration of the human brain. © SciePro, Shutterstock.com
p121 Vocal folds. The vocal cords open to let air pass through the larynx, into the trachea. The vocal folds are open when we breath in and closed when we want to speak. © Designua, Shutterstock.com
p123 The Digestive system with appendix. © Marochkina Anastasiia, Shutterstock.com
p124 The arch and branches of the aorta. © logika600, Shutterstock.com
p125 Neuron anatomy. Axon, myelin sheat, dendrites, cell body, nucleus. © LDarin, Shutterstock.com
p126 Cell Division under a microscope. © Rattiya Thongdumhyu, Shutterstock.com
p127 Ear anatomy. Cross section of External (outer), middle, and Inner ear opened. © Designua, Shutterstock.com
p128 Human skin. Layered epidermis with hair follicle, sweat and sebaceous glands. © Net Vector, Shutterstock.com
p129 The human digestive system. © medicalstocks, Shutterstock.com
p130 DNA helix break. © Anusorn Nakdee, Shutterstock.com
p131 Human eye close up. © H_Ko, Shutterstock.com
p132 Long bone structure. © sciencepics, Shutterstock.com
p133 Human eye anatomy. © Marochkina Anastasiia, Shutterstock.com
p135 Bones of the human hand. © studiovin, Shutterstock.com
p140 Fat cells adipocyte or lipocyte anatomy as an internal microscopic human cell for the storage of fats as a 3D render illustration. © Lightspring, Shutterstock.com
p140 3D model of the human spine. © CLIPAREA l Custom media, Shutterstock.com
Back cover Human heart illustration, Lightspring / Skeleton Hand, Big Pants Production / Lomitapide cholesterol lowering drug molecule, StudioMolekuul / white blood cells, Shilova Ekaterina © Shutterstock.com

# The Foundations of Science

*The Foundations of Science* introduces children to the wonders of the natural world in light of God's providential care over creation.

Too often we hear that science is in conflict with faith, but Pope St. John Paul II wrote that faith and science "each can draw the other into a wider world, a world in which both can flourish." *Foundations* seeks to spawn this flourishing in the hearts and minds of young readers, guiding them into a world that will delight their imaginations and inspire awe in the awesome power of God.

This eight-part series covers an extensive scope of scientific studies, from animals and plants, to the galaxies of outer space and the depths of the ocean, to cells and organisms, to the curiosities of chemistry and the marvels of our planet. Still more, it reveals the intricate order found beneath the surface of creation and chronicles many of the Church's contributions to science throughout history.

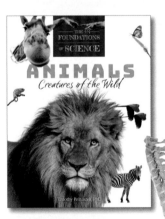

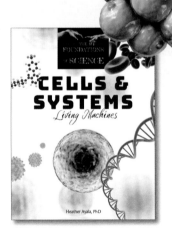
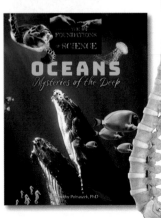
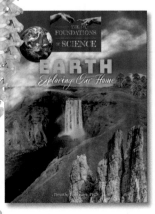
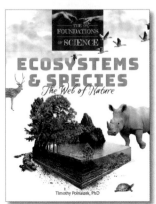
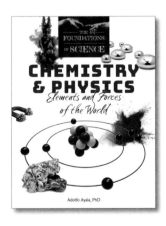